A Complete Guide to Wireless Sensor Networks

From Inception to Current Trends

A Complete Guide to Wireless Sensor Networks

From Inception to Current Trends

by

Ankur Dumka, Sandip K. Chaurasiya,
Arindam Biswas, and Hardwari Lal Mandoria

CRC Press
Taylor & Francis Group
Boca Raton London New York

CRC Press is an imprint of the
Taylor & Francis Group, an **informa** business

CRC Press
Taylor & Francis Group
6000 Broken Sound Parkway NW, Suite 300
Boca Raton, FL 33487-2742

First issued in paperback 2022

ISBN 13: 978-1-138-57828-9 (hbk)
ISBN 13: 978-1-03-240141-6 (pbk)

DOI: 10.1201/9780429286841

Publisher's Note
The publisher has gone to great lengths to ensure the quality of this reprint but points out that some imperfections in the original copies may be apparent.

**Visit the Taylor & Francis Web site at
http://www.taylorandfrancis.com**

**and the CRC Press Web site at
http://www.crcpress.com**

Contents

SECTION VII FAULT-TOLERANT WIRELESS SENSOR NETWORKS

SECTION VIII CROSS-LAYER OPTIMIZATION

About the Authors

Dr. Ankur Dumka is an Associate Professor in the Department of Computer Science and Engineering in Graphic Era (Deemed to be University), Uttarakhand, India. He has published more than 40 research papers of repute in international conferences and journals. He has written two books with international publishers and also contributed more than ten book chapters. He is associated with many reputed journals in the capacity of editor and editorial board member.

Sandip K. Chaurasiya received his M.E. degree in Computer Science and Technology from the Indian Institute of Engineering Science and Technology, Shibpur, Howrah, West Bengal, India. He is an assistant professor at the University of Petroleum and Energy Studies, Dehradun, India. His research interests include mobile computing, wireless communication, and wireless sensor networks.

Dr. Arindam Biswas received an M-Tech degree in Radio Physics and Electronics from the University of Calcutta, India, and a Ph.D. from NIT Durgapur, India. He was a postdoctoral researcher at Pusan National University, South Korea, with a prestigious BK21PLUS Fellowship. He was visiting professor at the Research Institute of Electronics, Shizouka University, Japan. Presently Dr. Biswas is working as an assistant professor in the School of Mines and Metallurgy at Kazi Nazrul University, Asansol, West Bengal, India. Dr. Biswas has nine years of experience in teaching, research, and administration. He has published more than 60 papers and authored five books. He has organized and chaired international conferences. His research interests are in carrier transport in low-dimensional systems and electronic devices, non-linear optical, communications, and bioinformatics.

Dr. Hardwari Lal Mandoria is a professor and the head of the department of information technology at G.B. Pant University of Agriculture and Technology, Pantnagar, India. He has 27 years of academic experience, has written over 100 research papers, and has published three book chapters. He has guided many doctorate and postgraduate research scholars under his supervision. He has been part of many advisory committees at the university level. He is associated with many journals and conferences in the capacity of editorial board member. He has also been part of Smart City, Dehradun, as academic expert committee member, contributing proposal drafting. His areas of interest include computer networks and information security.

INTRODUCTION AND APPLICATIONS

I

Chapter 1

An Introduction to Wireless Sensor Networks

Introduction

A sensor network consists of many sensing, computing, and communicating elements within a network which help an administrator to observe, react, and compute an event in a specific environment. These sensor networks can be used in a wide variety of applications in different fields like civil, commercial, and industrial. Sensor technology has evolved in recent years and has taken a central place within a wide variety of applications. Sensor technology can be used for collection of data, monitoring of data, surveillance of data, and other applications. Sensing devices can also be used in controlling devices by means of automation in sensing environments.

A sensor network can be categorized in four ways:

1. Assembly of localized sensors
2. Interconnecting networks
3. Clustering of information at central node
4. Data mining, data correlation, and data management through central node

Sensor nodes can accommodate large amounts of data that need certain algorithms for the management and processing of large amounts of data, and so this area of research involves the topic of sensor networks.

This chapter focuses on the fundamentals of wireless sensor networks (WSNs), their history, and other aspects related to them.

History of WSNs

Wireless sensor networks are evolving on a day-by-day basis and are different to sensor-based networking used five to six years ago. WSNs originated in military and heavy industrial applications. The first wireless sensor network was developed in the 1950s by the U.S. military for a sound surveillance system (SOSUS) for detection and tracking of Soviet submarines. This system used acoustic sensors-hydrophones- distributed in the Atlantic and Pacific oceans. Later, the National Oceanographic and Atmospheric Administration (NOAA) used this technology for detection of events in the ocean (Jindal, 2018).

The U.S. defense agency, Defense Advanced Research Project Agency (DARPA), in the 1980s developed distributed sensor networks for finding anomalies and challenges in its implementation. This research paved the path for academicians in the field of wireless sensor networks. Since then many advances in the field of wireless sensor networks have been achieved and recorded and are being used in multiple applications in various fields.

Architecture for WSNs

WSNs use a five-layered OSI architecture.

Application layer
Transport layer
Network layer
Data link layer
Physical layer

These five layers are controlled by three cross-layers for efficient working of WSNs . These three layers are power management, connection/mobility management, and task management layers.

> *Application layer*: The application layer is used for the management of traffic and provides software for various types of applications. These applications are used to send queries for obtaining the information from the system.

Transport layer: This layer provides internet work communication. This layer uses multiple protocols for providing reliability to the system and avoiding congestion within the network. As wireless sensor network is based on multihop transmission of data; hence, we generally do not use a TCP-based connection. UDP connection is more desirable in WSNs.

Network layer: The network layer supports various types of routing protocols. Routing protocols are used to maintain various aspects like power consumption, memory, reliability, and redundancy types of factors. Thus, working on routing protocols is one of the prominent areas of work for researchers in WSNs. Routing protocol can be based on factors like flat routing, hierarchical routing, event driven, query driven, or time driven based on the type of application in which the WSN is being deployed. To provide redundancy of data, WSNs use data aggregation or data fusion. Data aggregation combines data from different nodes and changes it into meaningful information to reduce the need for redundancy of data and thus saves energy. Data fusion is a more advanced concept of data aggregation which is used to remove noise from aggregated data (Manjeshwar and Agarwal, 2001, 2002; Wagner, 2004).

Data link layer: This layer provides reliability of data from point-to-point to point-to-multipoint. This layer also supports error control and multiplexing of data. This layer supports MAC addresses which are hardware addresses of nodes. This layer is used to provide higher reliability, low delay, higher efficiency, and throughput (Karlof et al., 2004).

Physical layer: This layer provides the interface for transmission of the data stream over the physical medium. This layer deals with the frequency of data transmission that includes selection of frequency, generation of carrier frequency for modulation, signal detection, and security.

Types of Sensors

The heart of wireless sensor networks is the type of sensors used, and that is based on the type of applications using the sensors. The individual nodes of the wireless sensor networks are often termed as motes. These sensors can broadly be divided into three groups as follows:

1. *Microelectromechanical systems (MEMS)*: This category includes magnetometers, acoustic sensors, pressure sensors, and pyroelectric effect sensors.
2. *CMOS-based sensors*: This category consists of humidity sensors, temperature sensors, and capacity proximity sensors.
3. *LED sensors*: This category includes proximity sensors and light-based sensors.

Based on demand and requirement, combinations of sensors can be deployed to create a network. For example, for environmental conditions, we can use combinations of MEMS, CMOS, and LED sensors for measuring environmental conditions like humidity, temperature, pressure, and light. On similar ground for tracking people or devices, we can use a combination of all sensors such as proximity sensor, pyroelectric, and acoustic sensors. Thus, these systems will be used for preparing a smart system using sensor-based technology which reduces human input for any system and makes the system intelligent and smart.

A wireless sensor network consists of the following components for setting communications among different sensors. These components are:

Microcontroller: This is used for processing data, performing tasks, and controlling the functionality of other components within a node.

Transceiver: The functionality of sending and receiving merges into this single device. The sensor node uses the industry, scientific, and medical (ISM) band which provides a free radio spectrum allocation and global availability. These radio frequency bands are used for communications among nodes of a wireless sensor network. A wireless sensor network uses license-free communication frequencies of 173 MHz, 433 MHz, 868 MHz, 915 MHz, and 2.4 GHz. The transceiver can be in different states of sending, receiving, idling, and sleeping. These are often termed as radios which are used for the transmission of data or signals (Chiara et al., 2009).

Sensors: Sensors are hardware devices that are triggered upon change in the physical state of a system and convert to measuring an analog signal. These analog signals are digitized by means of an analog to digital convertor, sending them to the controller for further processing. The sensors can be passive omni directional, passive narrow beam, or active sensors.

Passive omni-directional sensors are self-powered and require energy to amplify analog signals and have no notion of direction in measurement. Passive narrow beam sensors have a well-defined notion of direction when measuring (e.g., camera). Active sensors actively probe the environment (e.g., sonar or radar) (Fabbri et al., 2009).

Memory: Most of the microcontroller is on-chip memory for storage of information. The requirement for memory depends on the application. Memory chips are used for storing application-related data. The size of memory is constrained to a few kilobytes.

Battery/power source: Energy is one of the most important components of WSNs. The sensing nodes need power for sensing, communication, and processing of data. The power sources needed vary from 1.2 to 3.7 volts batteries. Much research is going in the direction of finding green energy for nodes. Power source can be AA, coin batteries, or solar panels.

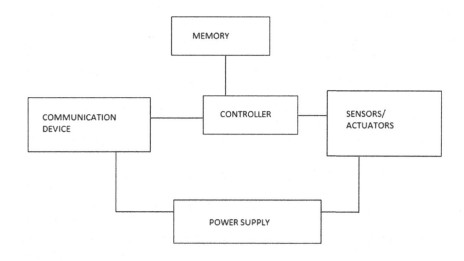

Types of WSNs

The type of wireless sensor network depends upon its use. WSNs can be deployed underwater, underground, or on land. Based on the area where it can be implemented, WSN types are specified as:

1. Underground
2. Underwater
3. Terrestrial
4. Multimedia
5. Mobile

In underground WSNs, nodes are deployed underground for monitoring conditions and sink nodes are placed above the ground to collect information from nodes under the ground and send it to the base station. Since the nodes are situated underground, the system is difficult and costly in terms of maintenance, deployment, and recharge. Attenuation and signal loss are also major issues in these WSNs.

Underwater WSNs deploy nodes under water for collection of information. The signals collected are sent to the sink node and from there to the base station for further processing. This network suffers with problems of long propagation delay, bandwidth, and faulty sensors due to sensors always being immersed in water. The charging or replacement of batteries is also an issue in these types of networks.

Terrestrial WSNs deploy sensors in the air which can be deployed in an ad hoc manner or structured manner depending upon the type of usage or application. Changing the battery is easy, and these networks are less expensive as compared to the other types of networks. These networks can make use of green energy by using solar. Energy conservation can also be achieved in these networks by means of minimizing delays, low duty cycle operations, and optimal routing.

Multimedia WSNs are used for tracking and monitoring of multimedia events like audio and video. This network makes use of low-cost sensor nodes which are equipped with cameras and microphones. The data is transmitted by interconnecting nodes. The data is processed by compression, retrieval, and correlation through the sensor nodes. This network faces major issues of high energy requirements, high bandwidth requirements, and innovative techniques for data compression and processing, as sensor nodes of this network have to carry multimedia that require high bandwidth and advanced processing techniques.

Mobile WSNs consist of mobile network nodes which can be moved from one place to another on their own to interact with the physical environment. These mobile nodes can compute input from the physical environment and communicate with neighboring nodes. This wireless network provides better coverage than other static networks and has better energy efficiency and channel capacity compared to static wireless networks.

Wireless Sensor Networks

A wireless sensor network is a self-configured and infrastructure-less wireless network that is used for monitoring of different physical and environmental conditions depending upon usage. The data of environmental conditions like temperature and humidity is sent from node to node to the mail location called the sink node, where the data is observed and analyzed. Sinks act as interfaces between user data and the network. Processed data can be retrieved from the sink node through queries. A WSN can consist of hundreds to thousands of nodes within a network, and these nodes communicate with each other by means of radio signals.

Designing wireless sensor networks requires different types of topologies for setting up nodes within a network. Some of the prominent topologies being used in WSN are star, mesh, and hybrid. The type of topology used for WSN depends on the type of usage and structure needed.

A wireless sensor network begins operation by using the sensing nodes to pick up data and collect information of interest. A sensor node consists of four basic components: sensing unit, processing unit, transceiver unit, and power unit. Any other additional unit can be connected or added based on the application type and use. The sensing unit is further subdivided into two sensor subunits and an analog to digital converter (Akyildiz et al., 2002). Sensors generate analog signals which are converted to digital signals by means of the analog to digital converter, and these are fed into the processing unit for further processing. The processing unit consists of a small memory and with a management procedure that carries out assigned sensing tasks by coordinating the sensing node with the other sensing nodes. The transceiver unit connects the nodes with the network to send out information. The power unit consists of power supplies which are used to provide power to the various components.

Sensing data from the source node to base station is done in a multihop environment which requires a routing path and routing protocols to make the routing paths. Different routing protocols have been proposed by different researchers (Akkaya et al., 2005; Akyildiz et al., 2002; Boukerche, 2009; Al-karaki et al., 2004; Pan et al., 2003; Waharte et al., 2006) that supports different criteria for forwarding of information. Different criteria for selection of routing protocols are optimized by time, resource, and power. There are three major areas for these protocols.

Issues and Challenges

Wireless sensor networks have changed ways of thinking and solving problems based on the types of applications, but there are many issues in the implementation of WSNs. One of the key challenges with WSN is that it uses a wireless medium for transferring data over the network. There are other challenges in energy consumption and the use of non-renewable forms of energy when transferring data from node to node. Many researchers have discussed the challenges of WSN (Akkaya et al., 2005; Akyildiz et al., 2002; SensorSim; Tossim et al., 2004; Pan et al., 2003). Some of these issues are:

Security: Security is one of the major issues with WSN. Zia and Zomaya (2006) have classified security of WSNs into four parts: confidentiality, integrity, authentication and availability. Sharma and Jena (2011) added one more to this security. Security in WSNs is needed needs to be seriously addressed.

Power consumption: One of the major issues with WSNs is power management, as all nodes need power to operate, and WSNs normally rely on non-renewable sources of energy. Another important aspect in terms of energy is the size of the batteries needed. Energy is also needed for data compression, which is again an issue. Limited energy of the nodes is a crucial issue, as it is impossible to replace or recharge batteries. Harvesting solar energy is not possible as it cannot be adopted due to the large size of solar panels. Power consumption in WSNs is an important factor equivalent to lack of bandwidth, memory, and processing capability. In order to increase the network lifetime, quality-of-service (QoS) support mechanisms need to be developed and should be lightweight with low and balanced energy consumption.

Fault tolerance: Node failure is one of the major issues with WSNs, as the nodes are frequently deployed in extreme environmental locations for gathering data, causing hardware problems, physical damage, and depletion of the battery. Deployment of protocols which can detect these failures at the earliest time is demanded for the faster retrieval of data and efficient and normal functioning of the network. The protocol should also vary by location and usage. The protocols used for fault tolerance are automatic repeat request (ARQ), forward error correction (FEC), and hybrid ARQ (Han et al., 2010).

Latency: Receiving exact information within a WSN is one of the primary aspects, and this can be achieved by minimizing delay by selecting a suitable routing protocol and network topology.

Throughput: WSNs use radio and packet transmission for sending communication from one node to another. The rate for successful transmission of data packets from source node to destination is called throughput, and the higher the throughput the more efficient the WSN.

Data suppression, aggregation, and fusion: Data aggregation combines data from different nodes and changes it to meaningful information to reduce the need for redundancy of data and thus saves energy. Data suppression, aggregation, and fusion minimize the data redundancy and reduce traffic load on a network, thereby ensuring energy savings. This elimination prevents redundancy, congestion, and overloading, decreasing the possibility of collisions, while improving bandwidth utilization. Data aggregation is a technique that eliminates duplicate packets and reduces power consumption, increasing WSN length of life.

Production cost: The frequent failure of sensor nodes may lead to a heavy cost for the WSN. The sensor node cost should be kept in mind before deployment of nodes in WSNs.

Scalability: WSNs can consist of hundreds to thousands of sensors based on the area of deployment and the type of deployment. For high resolution of data, node density may vary from place to place. The type of protocol being used for data retrieval needs to be scalable and such that it maintains an adequate performance (Intanagonwiwat et al., 2000).

Topology: Selection of topology should be based on the requirement to save energy. Topology maintenance is also an important aspect in terms of reducing energy consumption.

Transmission media: Radio communication over popular ISM bands is used for transmission of data from one node to another. There are some WSNs which might use optical-based communication or infrared-based communication, so the selection of transmission medium in such cases is of utmost importance.

References

C. Akkaya, J. Wiebe, and R. Mihalcea. Subjectivity word sense disambiguation. In *Proceedings of Conference on Empirical Methods in Natural Language Processing*, 190–199, 2009.

G. Al-Karaki, B. McMichael, B and J. Zak. Field response of wheat to arbuscular mycor-rhizal fungi and drought stress. *Mycorrhiza*. 14, 263–269, 2004.

A. Boukerche, *Algorithms and Protocols for Wireless, Mobile Ad Hoc Networks*. John Wiley & Sons, Inc., 2009.

B. Chiara, C. Andrea, D. Davide and V. Roberto, An overview on wireless sensor networks technology and evolution. *Sensors*, vol. 9, pp. 6869–6896, 2009.

F. Fabbri, C. Buratti, R. Verdone, J. Riihijarvi and P. Mahonen, Area throughput and energy consumption for clustered wireless sensor networks. In: *Proceedings of IEEE WCNC*, 2009.

X. Han, X. Cao, E. L. Lloyd and C. Shen, Fault-tolerant relay node placement in hetero-geneous wireless sensor networks. *IEEE Transaction on Mobile Computing*, vol. 9, no. 5, 2010.

C. Intanagonwiwat, R. Govindan and D. Estrin, Directed diffusion: A scalable and robust communication paradigm for sensor networks. In: *Proceedings of the 6th ACM 226 Routing Protocols for Wireless Sensor Networks, International Conference on Mobile Computing and Networking (MobiCom'00)*, 2000, pp. 56–67.

V. Jindal, History and architecture of wireless sensor networks for ubiquitous computing. *International Journal of Advanced Research in Computer Engineering & Technology*, vol. 7, no. 2, pp. 214–217, 2018, ISSN: 2278 – 1323.

C. Karlof, N. Shastry and D. Wagner, TinySec: A link layer security architecture for wire-less sensor networks. In: *SenSys'04*, 2004.

A. Manjeshwar and D. P. Agarwal, TEEN: A routing protocol for enhanced efficiency in wireless sensor networks. In: *1st International Workshop on Parallel and Distributed Computing Issues in Wireless Networks and Mobile Computing*, 2001.

A. Manjeshwar and D. P. Agarwal, APTEEN: A hybrid protocol for efficient routing and comprehensive information retrieval in wireless sensor networks. In: *Parallel and Distributed Processing Symposium, Proceedings International, IPDPS*, pp. 195–202, 2002.

J. Pan, Y. Hou, L. Cai, Y. Shi and S. X. Shen, Topology control for wireless sensor networks. In: *Proceedings of the 9th ACM International Conference on Mobile Computing and Networking*, 2003, pp. 286–29.

SensorSim [Online]. Available at: http://nesl.ee.ucla.edu/projects/sensorsim/.

S. Sharma and S. K. Jena, A survey on secure hierarchical routing protocols in wireless sen-sor networks. In: *ICCCS'11*, 2011.

D. Wagner, Resilient aggregation in sensor networks. In: *Proceedings of the 2nd ACM work-shop on Security of ad hoc and sensor networks*, 2004, pp. 78–87.

White paper by Silicon Valley "The Evolution of Wireless Sensor Networks."

O. Younis and S. Fahmy, HEED: A hybrid, energy-efficient, distributed clustering approach for ad hoc sensor networks. *IEEE Transactions on Mobile Computing*, vol. 3, no. 4, pp. 366–379, 2004.

T. Zia and A. Zomaya, Security issues in wireless sensor networks. *Systems and Networks Communications (ICSNC)*, pp. 40–40, 2006.

Chapter 2

Wireless Sensor Network Applications

Introduction

Real-life events become data by means of sensors, and data can then be processed, saved, or used in various applications. Sensors may be placed in differing environments based on how they are to be used. One example is the requirements for sensors used in the marine purpose (Albaladejo et al., 2010). The cascade of sensors should be designed such that they are waterproof and can withstand the effects of salinity and moisture. Similarly, sensors installed underground should have a high transmission power transceiver to overcome noisy channel attenuation due to being underground. The environmental changes noted by these sensors are communicated to the main server in order to make decisions in real time. Continuous monitoring of sensors means catastrophic failure can be prevented (Dargie and Poellabauer, 2010).

Solving technical issues of WSNs has widened the area of their applications. Issues like high power consumption have been improved with the reduction in the size of electronic devices so making it beneficial for large-scale remote installations. Wireless sensor networks sense the changes in the surrounding environment by means of sensor nodes, and through various methods relay the information collected to remote control centers for further action. In a wireless sensor network, there is a sink node that gathers data from all the other sensor nodes that collect information as per the requirements of the network. Sink nodes can have the same specifications as other sensors nodes or they can consist of customized devices like a PDA or laptop. They can be connected to other networks or consist of a base station that can be linked to network data by means of the internet to the remote control or monitoring center for the desired output.

With the flexibility provided by WSN technology in various fields of application, our way of thinking about life has been changed. WSNs have been successfully applied in various domains (Bharathidasan et al., 2001; Akyildiz et al., 2002, a,b; Sohraby et al., 2007; Yick et al., 2008; Boukerche, 2009; Buratti et al., 2009; Verdone et al., 2008) such as military, area monitoring, transportation, health, and environmental.

Applications and use of wireless sensor networks include intelligent monitoring of temperature, water level, humidity, pressure, and remote health monitoring of patients (Akyildiz et al., 2002; Yick et al., 2008). WSN deployment is dependable and can be modified based on the type of installation. Where there is a harsh environmental impact on sensors, the mechanical and electronic design should be robust. In the case of urban, rural, and suburban areas, the type of distortion in rural and suburban situations is lower as compared to the urban environment. This is due to increasing population density in urban areas that results in congestion and noise pollution. Due to continuous innovation in wireless sensor networks, the use of networks in unconventional applications has increased. Now a WSN can be used for monitoring sewage flooding, methane and other hazardous gases in sewers, disaster relief strategy, surveillance of criminal activities, and infrastructure monitoring (Lim et al., 2005; See et al., 2010).

Literature Review

Losilla et al. (2007) in their research have implemented WSNs for transportation purposes. Yang et al. (2007) deployed the WSN for a power delivery system–based application in order to measure sag in overhead conductors within the entire transmission line. He et al. (2004) integrated wireless sensor network technology for use in surveillance missions. This integrated system acquires and verifies information on enemy capabilities and positioning of hostile targets. It deploys an unmanned surveillance system using wireless sensor networks which have proven to be beneficial in military operations. Moturu et al. (2011) deal with the issue of improving the performance of wireless sensor networks by means of association control, and this is achieved by means of associating intelligent users with the network's access points. This provides an innovative and optimal solution for utilization of the reinforcement learning (RL) algorithm which is known as Gaussian processes temporal differences (GPTD). Wang et al. (2010) discuss supervised learning within wireless sensor networks. In their research they propose a new approach towards routing, optimization, and reliability. Routing is used in order to maintain information about neighbor states and other factors. Wang et al. focus on challenges in performing accurate and adaptive information discovery and also processing and analysis of extracted data for use in features such as co-relations. In order to achieve this, the authors propose use of supervised learning techniques for making informed decisions within wireless sensor networks.

Pawa et al. (2011) classify the routing protocols of wireless sensor networks into flat-based, hierarchical, and location-based. Low energy adaptive clustering hierarchy (LEACH) is an example of a hierarchical routing protocol which is energy efficient and works within parameters like network lifetime and stability period. It uses random rotation of nodes to find the cluster head within a cluster of wireless sensors so that energy can be evenly distributed.

WSN Applications

We can say that the range of applications for wireless sensor networks is very wide and broad, ranging from environmental to industrial to medical to military to habitat monitoring and many others, as can be seen in the following:

Area of application of wireless sensor network	Use
Asset management	Tracking of shipping containers
Air traffic control	Controlling air traffic pattern
Home automation	Multiple home system controls like conservation, convenience, safety, electric, water, and gas utility usage data, Home security
Military	Battlefield management, battlefield reconnaissance, surveillance, management, combat field surveillance, detecting structural faults in aircraft, detecting structural faults in ships, detection of enemy vehicles
Electricity management	Automatic meter reading, smart grid management, electricity load management
Biological field	Biological monitoring for agents, detecting toxic agents.
Medical	Biomedical applications, smart ambulance, real time monitoring of patients, wireless body area networks, collecting clinical data, heart beat sensor, telemedicine
Road safety and management	Bridge and highway monitoring and management

Construction	Building and structure monitoring, building automation, building energy control and monitoring, detecting structural faults in buildings
Disaster management	Earthquake detection, tsunami detection and response, disaster emergency response
Business	E-money applications, kiosks, monitoring and controlling workspaces, intruder detection
Habitat	Habitat monitoring, sensing
Industry	Industrial and building monitoring and automation, manufacturing monitoring and automation, asset management, process control, inventory management, manufacturing control, material processing systems
Power system	Monitoring of electrical distribution systems, smart grid sensor and actor network application, automated remote meter reading, thermal rating monitoring of conductors in power systems, monitoring sag clearance in overhead conductors
Transportation	Traffic monitoring, transportation-based application, vehicular ad hoc network (VANET)-based application, road-based application, car parking, underground railway tunnel monitoring,
Gas monitoring	Sewage gas monitoring, gas pipeline monitoring, gas meter monitoring, air pollution monitoring
Others	Commercial applications, consumer applications, consumer electronics and entertainments, tracking of belongings like pets, heating control

Initially wireless sensor network usage was only meant for military applications, but with advancements in wireless sensor technology areas of application have expanded.

Medical: WSNs can be used for supporting interfaces for the disabled. They can also be used for monitoring and diagnosing patients in real time. WSN technology can be used for drug administration in hospitals. One application of WSNs in medical care is tele-monitoring of physiological data.

Another application is the smart ambulance (Dumka, 2018) which will enable doctors to monitor and start their procedures for treatment in the ambulance. The smart ambulance allows for heartbeat monitoring and other medical procedures.

Power system: WSNs can be used for setting up a smart grid that is helpful in providing low-cost, flexible, low-power dissipation, and self-organized power. WSN technology can be useful in power systems for smart metering, distributed bus protection of power networks, and fault location.

Smart grid is one of the major contributions to the field of power systems that can be used with non-conventional and conventional sources of energy. WSN technology can be used for remote monitoring and controlling of a smart grid and can be used for maintenance and avoidance of major faults (Dumka, 2017). A power sensing module can be designed for calculating the power of any kind of load. The information can be communicated to the sink on regular intervals, where the data can be processed and used for deciding on suitable actions like detection of power theft, energy efficient building design, smart metering, and smart automation.

Geo sensing: With the advancement of technology and the introduction of IoT technology, the requirement for long-range wireless networks has increased. WSNs can be used to get information on a moving object and send over the network. This information is collected by GPS and LoRA (long-range) modules.

WSN technology can be used to track animals, humans, objects, and vehicles in order to get real-time data on these objects. This information can be sent to a remote base station by means of WSN technology, and the data can be processed by means of certain other tools and technologies for taking appropriate decision on such data.

Water Pipeline Monitoring

Pipeline leakage is a major issue in water wastage. WSN technology supplies a solution in this respect by decreasing leakage and conserving water. The WSN solution can be used to detect pipeline leakage and can also be used for monitoring leakage within a large pipeline system. Several leak detection and localization algorithms can be used with an efficient wireless sensor node system on chip (SoC) used in real time. Leaks can be detected using fluid mechanics and kinematics physics based on the harnessed water flow rate obtained using flow liquid meter sensors and microcontrollers.

The use of this technology means human patrolling along the pipes looking for visual leaks and use of ultrasound or acoustics equipment to search for the leaks is no longer needed. Water pipeline monitoring systems use sensors to collect data and analytics tools and techniques for processing data in an efficient manner. There are various methods provided by various researchers, such as Kim et al. (2016), in this direction as non-intrusive, autonomous water monitoring system (NAWMS).

Pipe probe, proposed by Chang et al. (2011), describes a prototype for a mobile sensor network system with a hydro molecular form. This prototype uses pressure sensors MS5541C and EcoMote. It gathers measurement from the pipe and saves information in flash memory where the readings are analyzed and suitable solution provided that require human interactions.

Sensor-based pipeline autonomous monitoring and maintenance system (SPAMMS) was proposed by Kim et al. (2016) and uses static and mobile sensors for leakage detection within the pipeline system. Pressure sensors, chemical sensors, CCD sensors, and sonar sensors can be used. This system uses robot technology for maintenance and monitoring of a pipeline system. The robot technology consists of a MiCA1 mobile sensor mote, an EM 4001 ISO radio-frequency identification (RFID), and a robot agent. This system uses high processing technologies like image processing and signal processing algorithms.

Solid Waste Management

Waste management is one of the biggest problems in keeping a city or a country clean and disease free. The waste management cycle includes waste generation by industry, markets, and houses in which garbage is collected by municipal corporations that dump it in landfill sites.

Solid waste management using WSNs focuses on waste management automation where waste bins are attached with proximity sensors that trigger an event as and when the bins are full of garbage. This full bins information can be sent to the municipal authorities by means of WSN technology, enabling them to reach a particular bin position as soon as it is full. Thus, WSN-based solid waste management provides an automated solution enabling municipal authorities to have immediate knowledge of full bins so that they can take empty the bins in a timely fashion.

There are several researchers who propose some solutions for WSN solid waste management. Chaudhary et al. (2011) in their paper propose an RFID and load cell sensor–based waste management system for detection of waste in real time. Hanan et al. (2012) in their paper propose a model with a framework consisting of RFID and communicating devices like GSM, GIS, and GPRS for garbage monitoring and management activities.

Temperature Monitoring

A WSN can be used for monitoring multiple environmental conditions such as temperature. A WSN infrastructure can be used for setting up communication among sensors. These sensors can be used to gather information and data on various environmental parameters like temperature and humidity.

Different researchers have adopted different approaches to extracting temperature data from the environment and performing analytics. Bin et al. (2011) used WSN-based Zigbee technology for extracting temperature data using thermocouples as temperature sensors. Peng and Wan (2013) used infrared-based temperature sensors for extracting data from environmental variables and this data is transferred to a system by means of RS232. Mainwaring et al. (2002) used WSN technology for real-world habitat monitoring to extract environmental data including temperature.

Structural Monitoring

WSN technology can be used in civil work such as buildings and infrastructure. WSN enabled engineering practices can be applied to bridges, flyovers, tunnels, and embankments for monitoring progress of work being done without having to visit the site. Thus, WSN can be used for remote monitoring of any civil work.

Air Pollution Monitoring

Rapid urbanization and industrialization degrade environmental quality parameters. Tracking of environmental indices in order to develop realistic models and take appropriate decisions. There are many WSN-based air quality monitoring systems being proposed by many scientists to track and fix increased air pollution levels.

There are a number of sensors that can be used for monitoring environmental conditions. Some examples are MQ-7 and MQ-2 sensors which can be used for monitoring carbon dioxide and carbon monoxide levels in the atmosphere. There are certain ARM processors such as LPC 2148 that can be used for interfacing with these sensors. Zigbee technology like Tarang F4 can be used for wireless processing of data according to conditions. Applications can be developed to retrieve data as and when required. The data retrieved can be used to make decisions for taking suitable action to prevent further pollution.

Gas Monitoring

Monitoring gases is an important use for a WSN. WSN technology by means of sensors can detect gas leaks within a system, and the same information can be sent to any remote location user by means of WSN technology or integrated with IoT can be sent to any user on their phone. Thus, any user can get information for a gas leak in their home, shops, or any other place on a real-time basis and take appropriate action.

WSN technology can be used in different applications for gas monitoring:

1. Gas meter monitoring
2. Gas pipeline monitoring
3. Sewage gas monitoring

Intruder Detection

A WSN can be used for intruder detection. Sensors are used to detect intruders within a system and send notification by means of WSN technology to a controller and then raise an alarm. Intruder detection system can be of two types: rule-based intruder or anomaly-based.

The rule-based intruder detection system is based on a signature type which works via a built in signature. This detection system has high detection rates for already known types of attacks, but does not work as well for any new types of attacks. The anomaly-based intruder detection system works by matching traffic patterns or resource utilizations. This type of intruder detection system works well for new types of attacks.

Disaster Management System

A wireless sensor network can be used for early detection of disasters in certain land masses so that people can take action. This WSN consists of sensors which detect any incoming disaster events and communicates with the base station to send data to the main station. Sensors sense the environmental surroundings within parameters like temperature, pressure, and frequency, and they generate a signal accordingly. This signal is compared with threshold values of that normal attribute, and if it exceeds the threshold value, then the signal is transmitted to the base station from where it goes to the appropriate authority for taking necessary action. Messaging can be used to send information to every individual in the case of an earthquake or tsunami type of emergency.

A disaster management system can be used for:

1. Emergency response system
2. Tsunami detection and response system
3. Disaster surveillance

Transportation

There are many applications for WSNs in transportation systems. WSN technology can be used for detection of traffic at any point in the system and can send early warning messages to all incoming traffic to that location to take a different route. Various types of sensors such as proximity and image can be used for detecting traffic. These can detect the number of vehicles at a certain point and send signals to the base station where this number is compared to the threshold value. If the number exceeds the threshold value, this triggers a message to all other incoming vehicles, preventing heavy traffic and congestion (Dumka, 2018).

Various WSN technology has been developed in recent years like mobile ad hoc networks (MANETs) and VANETs that have made for tremendous changes in traffic patterns and changed the way of thinking about using WSN in transportation systems. Some ways to using WSNs:

1. Traffic monitoring
2. Transportation-based application

3. VANET-based application
4. Road side assistance

Power System

Power system problems include power theft, power outage, and grid failure. WSNs have provided a major relief for these types of problems by providing real-time-based applications that detect power usage patterns, and this data in combination with concepts like big data can provide solutions.

Smart meters use sensor-based technology that senses power usage in real time, sending that information to a near base station that then sends the information to a main base station. The data can be processed and managed by means of a big data application and can provide solutions such as the real-time value of meter reading, real-time data based on location, user-based segmentation, area-based segmentation, and usage-based segmentation (Dumka, 2018). Some of the applications of WSN in power systems are as follows:

1. Electrical distribution system
2. Smart grid
3. Remote meter reading

Home Control

A wireless sensor network can be used for the controlling, conservation, convenience, and safety of home appliances and applications. A wireless sensor network can be used for heating, lighting, and cooling a home from a remote location. Wireless sensor networks can be used for controlling, various home systems. They can be used for capturing data on electric, water, and gas utilities. In combination with embedded intelligence, wireless sensor networks can be used for optimizing the consumption of natural resources. Thus, by means of wireless sensor technology we can remotely control all home appliance and applications. This technology will also enable users at remote location to detect or receive notifications of any unusual events occurring in the home.

Building Automation

Wireless network technology with technology like Zigbee can be used for controllable light switches, resulting in energy savings. Energy management in hotels is also an application. Centralized HVAC management saves energy by turning off cooling in empty rooms. Thus, using wireless sensor network technology, we can integrate and centralize management of lighting, cooling, heating, and security, controlling multiple systems from a centralized system, reducing energy by means of HVAC management, reconfiguring lighting systems for adaptable workspaces, upgrading building infrastructure with minimal effort, integrating and storing data from different sensors for analysis and future use.

Industrial Automation

Industrial automation can be used for the controlling, conservation, safety, and efficiency of industrial equipment in multiple ways. Automation by means of wireless sensor networks can be used for automation of manufacturing and process control systems, improving asset management, identification of faulty operations in real time, acquisition of data through remote sensors to reduce human intervention, and monitoring of networks for enhancing safety of the public and employees. Industrial automation can use RFID technology, a technology that increased world industrial revenue by $2.8 billion in 2009. In 2009 many industries adopted the technology to make their business processes more efficient.

Military Applications

Wireless sensor technology can be used in many military applications like surveillance and border monitoring. There are many new technologies and equipment developed by many companies which provide suitable military applications. Being smaller in size and having more capabilities in terms of features like robustness, self-organization, and networking than existing systems, make WSNs a better solution for surveillance in the military. WSNs by means of distributed sensing technology can be used for redundant and reliable information on threats and for localizing threats among distributed sensor nodes. WSNs can be used for monitoring large areas and perimeters to achieve goals of self-protection.

Habitat Monitoring

Many projects have been implemented for protecting and monitoring habitat. One example is Intel Research Laboratory in Berkeley in collaboration with the College of the Atlantic in Bar Harbor who set up a project that deploys a wireless sensor network on Great Duck Island in Maine. This project aims to monitor microclimates in and around nesting burrows used by Leach's storm petrel. This project's aim is to develop a habitat-monitoring kit which can be used by researchers worldwide for non-intrusive and non-disruptive monitoring of sensitive wildlife and habitat (Intel Research Laboratory). In order to achieve this, three dozen nodes are deployed on the island. These nodes monitor the nesting habits of the Leach storm petrel and send the data to the satellite link which allows retrieval of data from anywhere in the world. Habitat monitoring can also be achieved by means of taking data on temperature, humidity, and pressure so that appropriate decisions can be made for habitat preservation and prevention.

Conclusion

This chapter reviewed different applications of WSN that focus on environmental variables, disaster systems, power systems, habitat monitoring, power systems, and others. The thrust of this chapter is to open up an area of thinking of applications of WSNs in different areas, as these can be unlimited.

References

I. F. Akyildiz, W. Su, Y. Sankarasubramaniam and E. Cayirci, A survey on sensor networks. *IEEE Communications Magazine*, vol. 40, pp. 102–114, 2002a.

I. F. Akyildiz, W. Su, Y. Sankarasubramaniam and E. Cayirci, Wireless sensor networks: A survey. *Computer Networks*, vol. 38, pp. 393–422, 2002b.

C. Albaladejo, P. Sánchez, A. Iborra, F. Soto, J. A. López and R. Torres, Wireless sensor networks for oceanographic monitoring: A systematic review. *Sensors*, vol. 7, pp. 6948–6968, 2010.

A. Bharathidasan, V. Anand and S. Ponduru, Sensor Networks: An Overview, Technical Report. Department of Computer Science, University of California, Davis, 2001.

C. Bin, J. Xinchao, Y. Shaomin, Y. Jianxu, Z. Xibin and Z. Guowei, Application research on temperature WSN nodes in switchgear assemblies based on TinyOS and Zigbee. In: *4th International Conference on Electric Utility Deregulation and Restructuring and Power Technologies (DRPT)*, 2011, pp. 535–538.

A. Boukerche, *Algorithms and Protocols for Wireless, Mobile Ad Hoc Networks*. John Wiley & Sons, Inc., 2009.

C. Buratti, A. Conti, D. Dardari and R. Verdone, An overview on wireless sensor networks technology and evolution. *Sensors*, vol. 9, pp. 6869–6896, 2009.

D. T. T. Chang, Y. S. Tsai, H. C. Lee, L. L. Guo and K. C. Yang, Wireless sensor network (WSN) using in tunnel environmental monitoring for disaster prevention. *Advanced Materials Research*, vol. 291–294, pp. 3401–3404, 2011.

D. D. Chaudhary, S. P. Nayse and L. M. Waghmare, Application of wireless sensor network for green house parameter control in precision agriculture. *International Journal of Wireless & Mobile Networks (IJWMN)*, vol. 3, pp. 140–149, 2011.

W. Dargie and C. Poellabauer, *Fundamentals of Wireless Sensor Networks: Theory and Practice*. John Wiley & Sons, Hoboken NJ, 2010.

A. Dumka, *Computational Intelligence Application in Business Intelligence and Big Data Analytics*. CRC Press, Boca Raton, FL, 2017.

A. Dumka, IoT based traffic management tool with Hadoop based management scheme for efficient traffic management. *International Journal of Knowledge Engineering and Data Mining*, vol. 5, no. 3, pp. 208–221, 2018.

M. A. Hannan, A. Mustapha, A. Al Mamun, A. Hussain and H. Basri, RFID and communication technologies for an intelligent bus monitoring and management system. *Turkish Journal of Electrical Engineering and Computer Sciences*, vol. 22, 2012.

T. He, S. Krishnamurthy, J. A. Stankovic, T. Abdelzaher, L. Luo, R. Stoleru, T. Yan, L. Gu, J. Hui and B. Krogh, Energy-efficient surveillance system using wireless sensor networks. In: *MobiSys 2004 – Second International Conference on Mobile Systems, Applications and Services*, 2004.

S. Kim, Y. Yim, S. Oh and S. H. Kim, Social wireless sensor network toward device-to-device interactive Internet of Things services. *International Journal of Distributed Sensor Networks*, vol. 12, 2016.

H. B. Lim, Y. M. Teo, P. Mukherjee, V. T. Lam, W. F. Wong and S. See, Sensor grid: Integration of wireless sensor networks and the grid, 2011, pp. 91–99.

F. Losilla, C. Vicente-Chicote, B. Álvarez, A. Iborra and P. Sanchez, Wireless sensor network application development: An architecture-centric MDE approach, In: *European Conference on Software Architecture*, 2007, pp. 179–194.

A. Mainwaring, J. Polastre, R. Szewczyk, D. Culler and J. Anderson, Wireless sensor networks for habitat monitoring. In: *Proceedings of the ACM International Workshop on Wireless Sensor Networks and Applications*, 2002.

S. T. Moturu, I. Khayal, N. Aharony, W. Pan and A. Pentland, Using social sensing to understand the links between sleep, mood, and sociability. In: *2011 IEEE Third International Conference on Privacy, Security, Risk and Trust (Passat) and 2011 IEEE Third International Conference on Social Computing (Socialcom)*, 2011, pp. 208–214.

T. D. S. Pawa and S. K. Jena, Analysis of low energy adaptive clustering hierarchy (Leach) protocol, National Institute of technology Rourkela, B.Tech Thesis, 2011.

D. Peng and S. Wan, Industrial temperature monitoring system design based on Zigbee and infrared temperature sensing. *Optics and Photonics Journal*, vol. 3, no. 2B, pp. 277–280, 2013.

C. H. See, K. V. Horoshenkov, S. J. Tait, R. A. Abd-Alhameed, Y. F. Hu, E. El-Khazmi, J. G. Gardiner, A Zigbee based wireless sensor network for sewerage monitoring. In: APMC—Asia Pacific Microwave Conference , 2010, pp. 731–734.

K. Sohraby, D. Minoli, T. Znati, *Wireless Sensor Networks: Technology, Protocols, and Applications*. John Wiley & Sons, Hoboken, NJ, 2007.

R. Verdone, D. Dardari, G. Mazzini and A. Conti, Wireless sensor and actuator networks. *Wireless Sensor and Actuator Networks*, 2008.

Y. Wang, B. Ramamurthy, X. Zou and Y. Xue, An efficient scheme for removing compromised sensor nodes from wireless sensor networks. *Security and Communication Networks*, vol. 3, pp. 320–333, 2010.

W. Yang and Y. Huang, Wireless sensor network based coal mine wireless and integrated security monitoring information system. *Sixth International Conference on Networking*, 2007, p. 13.

Y. Yang, A. Mohammat, J. Feng, R. Zhou and J. Fang, Storage, patterns and environmental controls of soil organic carbon in China. *Biogeochemistry*, vol. 84, pp. 131–141, 2007.

J. Yick, B. Mukherjee and D. Ghosal, Wireless sensor network survey. *Computer Networks*, vol. 52, pp. 2292–2330, 2008.

PHYSICAL
SUPPORT LAYER

Chapter 3

Quality of Service-Sensitive MAC Protocols in Wireless Sensor Networks

Introduction

Wireless sensor networks integrate automated sensing and embedded devices into wireless networking and are an emerging and evolving technology. Wireless research has moved on from agriculture (Wark et al., 2007) and environmental monitoring (Mainwaring et al., 2002) to more complex applications such as industrial monitoring, automation (Gungor et al., 2009), and health care. Wireless sensor network technology is developing more rapidly due to the invention and availability of low-cost hardware, tiny cameras, and microphones. These have enabled multimedia and visual wireless sensor network technologies for applications such as surveillance. These newer applications require quality of service (QoS) for their best effort performance.

With the demand for bandwidth and competition in the market, quality of service (QoS) has become a benchmark in wireless sensor networks. QoS has been defined by the International Telecommunication Union (ITU) as the "totality of characteristics of a telecommunication service that bear on its ability to satisfy stated and implied needs of the user of the service." Earlier QoS was about conserving resources rather than providing service quality. QoS is now used for assigning various priorities for users, applications, and data flow to ensure efficient and effective service implementation. QoS can be implemented by means of controlling resource sharing to ensure a higher performance level by means of setting various measurable parameters like jitter, delay, packet loss, and available bandwidth.

Early QoS implementation to provide efficient service relied on end-to-end signaling for applications like multimedia. For this approach a reservation-based approach like integrated services or IntServ (Braden et al., 1984) was used. The IntServ approach is used to provide guaranteed QoS service within the network, which is always a challenge due to the unpredictable nature of wireless links, resource constraints within a WSN, and unstable topology that may occur due to node or link failure. With the advancement of technology and the need for more bandwidth, the requirements for QoS support within a WSN are needed even more. There is a need for well-designed and novel QoS support in each layer of a communication's protocol stack to ensure sensitivity to bandwidth and delay in WSNs. Real-time and mission-critical applications demand more stringent QoS requirements for reliable and delay-bound data transfers.

In this chapter, we focus on support of QoS in the MAC layer of WSN architecture. The chapter discusses distributed QoS support in the MAC layer because all other layers of a WSN architecture depend on the MAC layer, making the MAC layer a decisive factor in the overall performance of the network. Recently, a cross-layer solution has been implemented that combines the functionality of multiple traditional layers into functional modules (Melodia et al., 2006). The cross-layer solution avoids layer abstraction within the protocol stack. While in the case of QoS, there is no difference between layered and cross-layered protocols. This chapter presents studies of existing QoS-aware MAC protocols, including underground, underwater, and mobile sensor networks.

Zogovic et al. (2010) in their research have focused on provision of QoS at the MAC and physical layers. The authors have shown the classification of QoS at MAC with a direction for future research without making any comparison. This chapter covers their detailed survey of QoS at the MAC layer, introducing background information on provision of QoS in wired and wireless network and discusses requirements for different types of applications. We have divided QoS provision into two parts as seen below:

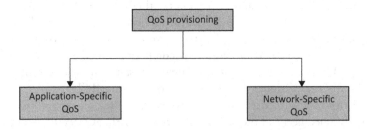

This chapter discusses both types of QoS provisioning and their challenges for providing QoS within network parameters like delay, throughput, duty cycle, and contention window at the MAC layer.

An Overview of QoS

The internet was originally designed to provide best effort delivery of services for application data (Tsigkas et al., 2008). With research and the emergence of new application types such as video, telephony, and IPTV that demand high throughput, high reliability, bounded jitter delay, and best effort service delivery, it has become more difficult to support these sorts of applications. Such applications need the support of algorithms, protocols, and mechanisms which provide QoS. WSN is another area which requires algorithms for different applications with an expansion in the implementation of WSNs in more areas. WSNs traditionally were used for low-rate data collections, along with low periods of operations for monitoring purposes. With the increase in use of WSNs in complex operations such as assisted living (Wood et al., 2008), target tracking (He et al., 2006), and water monitoring, the need for timely, reliable, and efficient data collection is more crucial. With recent advancements in the use of image sensors which enable usage of video sensors, there has evolved a new class of WSNs called visual or multimedia sensor networks (Akyildiz et al., 2007; Soro et al., 2009) that can be used for applications such as surveillance. With an increase in the amount of data, the need for QoS also increases.

QoS meets the demand of the user and/or applications. QoS provisioning can be categorized as soft QoS or hard QoS depending upon its application in wireless or wired networks. In hard wired QoS, there is strict bounding of factors affecting QoS guarantees such as packet loss, packet delay, and bandwidth. In soft QoS there is some flexibility on these factors (Tsigkas et al., 2008).

Two types of service differentiation model can be adapted for wired and wireless networks for providing efficient QoS. These are differentiated services (DiffServ) and Intserv (Dumka, 2018). Both these services are used for prioritizing packets. Priorities are defined by their service quality and thus improvement of service is achieved by sharing available resources among them. Diffserv focusses on packets for providing efficient QoS, whereas Intserv focusses on packet flow for providing efficient QoS (Table 3.1).

Scope of QoS in a WSN

QoS scope in a WSN has been described by different authors from different perspectives. Chen et al. classified QoS into two as application-specific and network-specific.

Application-specific focuses on the quality of the application. These specifications include deployment of the application, lifetime of the application (Slama et al., 2007; Xiao et al., 2006) coverage of the application (Gu et al., 2008), and the number of active sensors (Ma et al., 2008; Younis et al., 2004). Based on the demand for QoS from the customer, an application-specific QoS can be designed.

Table 3.1   Relationship between Different Services

Differentiated Services	Integrated Services
In the differentiated service model, a packet "class" can be marked directly in the packet.	In the integrated service model, special QoS treatment to the flow of router is determined by signaling.
Differentiated service achieves better QoS	Integrated service QoS is used for real-time traffic
DiffServ model is RFC-2475	IntServ model is RFC-1633
Resources are allocated by dividing the traffic into small number of classes by the DiffServ architecture	Real-time applications like remote video, multimedia conferencing, real-time applications are motivated by IntServ architecture
No end-to-end signaling	It provides end-to-end QoS
"Admission control" and "resource reservation" are not supported by DiffServ	For establishing QoS, "resource reservation" and "admission control" are supported by IntServ. RSVP is used to signal explicitly the need for QoS of an application's traffic in end-to-end path through network

A network-specific perspective provides QoS service during delivery of data within a network. It provides efficiency for the network by focusing and managing latency, reliability, and packet loss in each layer.

One of the severe challenges in imposing QoS within a WSN can be harsh environmental conditions and severe resource constraints. One limited resources constraint in a WSN is node battery issues. Limited energy support from the batteries in WSN nodes is a problem, and it is a recurring issue of replacing and recharging batteries. Running WSN nodes on solar energy (Corke et al., 2007; Taneja et al., 2008) is one solution for energy scarcity, but large solar panels are still a problem for tiny sensor nodes. A QoS mechanism needs to be proposed for this problem and it should be light weight and simple, operable on highly resource constrained sensor nodes.

Node deployment can be either random or deterministic based on the type of network being deployed. In the case of deterministic deployment, sensor nodes are deployed by hand and predetermined scheduled paths have been set up for the purpose of routing. In random node deployment, the nodes are randomly arranged and organized in an ad hoc manner. So in this type of deployment, path discovery, neighbor discovery, geographical information of nodes, and clustering are few of the issues that need to be discussed and managed using QoS.

The topology of a WSN can change due to natural events like fire and flood which can lead to problems like link failure, node mobility, malfunctioning of nodes, and energy depletion. MAC layer protocol uses a sleep-listen schedule and to save energy by temporarily turning off the radio of the sensor. Such plans are also a cause of frequent changes in topology. This topology change also requires implementation of QoS to overcome it.

Various sensor nodes are deployed in a WSN where an observed event or phenomena can be detected by more than one node and can create redundancy of data within the network. This redundancy can cause unnecessary data delivery in the network causing network to be flooded with the same information. Data aggregation and fusion techniques proposed by Krishnamachari et al. (2004) and Intanagonwiwat et al. (2003) can be used to prevent redundancy of data, but further QoS solutions are to reduce additional delay and complexity within a system.

There are cases when a single node can be used for capturing different attributes of different types of traffic within an area. QoS support in terms of differentiating these classes is required.

QoS is also required for real-time data such as weather forecasting, emitted gases data, and disaster monitoring. Data in such cases is limited to specific time frames and there needs to be instant data transfer inthe time frame. This critical time parameter means real-time applications need better QoS.

There are many other parameters that required QoS for better implementation of a network like unbalanced traffic patterns and scalability of the network. These demand QoS service for better and efficient transmission of data from node to node.

The introduction of mobile sensor nodes and their use in a number of applications also increases the demand for QoS within a WSN. These mobile nodes set a challenge for management of topology, decisions on routing, and energy management. Due to the mobility of nodes, the topology keeps changing and also the spatial density of the nodes within the network changes.

The following challenges discussed are some of the issues that have been faced by WSNs. These challenges make it difficult for implementation of QoS and providing deterministic QoS guarantees in WSNs. This chapter discuss different types of QoS provision within a WSN for efficient transmission of packets.

QoS Requirements, Metrics, and Parameters

QoS requirements differ between one application and another. The QoS requirement is discussed based on the basic data delivery models:

1. *Event-driven model*: In this model the sensors will activate and sense data in only in the case of an event. In this model, when an event occurs, a burst packet is generated that reports the event to the base station. The efficiency of

the model depends on how fast and accurately the sensors act on the event in sending information to the base station. The sensed information is sent from node to node to reach the base station. Examples of such models are surveillance systems.

2. *Query-driven model:* The query driven model is similar to the event driven model with the difference being that data is requested by the sink whereas in the event driven model data is pushed to the sink without any demand. Query driven supports two-way traffic, whereas event driven only supports one way traffic. The query driven model supports quick and reliable node request and response to achieve a better network performance. An examples of the query driven model is habitat monitoring.

3. *Continuous model:* In this model, nodes are continuously transmitting the collected data to the base station. To store scalar data and to save energy, the nodes are activated only during data collection. Depending upon the type of data, the bandwidth requirements changes. An example is saving data of video and voice where more bandwidth is required for continuous data transmission to the base station.

4. *Hybrid model:* If the aforementioned data delivery model is a combination of more than one model, then this type of model is termed a hybrid model.

QoS Metrics

There are various QoS metrics which are need for management of providing efficient service. Some of the major QoS metrics parameters are bandwidth, delay, reliability, jitter, and energy consumption. In order to provide the best QoS, the network should provide maximum throughput, reliability, and energy efficiency and minimize delay and jitter. In the MAC layer, QoS parameters affecting network service are as follows.

For minimizing the end-to-end delay within the MAC layer, it is necessary to minimize the medium access delay of the sensing device to ensure minimal packet latency for end-to-end packet delivery.

In order to prevent collision within the MAC, a layer carrier sensing method can be used. As per the traffic requirements, contention window protocols can be adopted. Collisions can be reduced by contributing to reliability assurance. Packet loss can be detected using acknowledgement, and retransmission can be performed in time.

Loss of energy is one of the important factors in a WSN due to the limited operation capacity of batteries in sensor nodes. Energy dissipation can be reduced in the MAC layer by minimizing collision and retransmission and tuning the duty cycle of the sensor nodes according to network dynamics by turning off nodes when not in use.

One of the key components of WSNs is change adaptation. This factor is crucial in the case of WSN due to the dynamic behavior of WSNs. Adding new nodes and changing nodes due to environmental factors or topological changes are some of the common issues in WSN. There is a need for adaptive actions at the MAC layer for such types of situations. Interference and concurrency are also major metrics needed in network deployment. There are other parameters which play important roles in WSN like bandwidth, acknowledgement mechanism, security, transmission power, and frequency of transmission.

It is not mandatory or practical to provide each MAC-related QoS metric in a single MAC protocol, as application of WSNs differ. A single MAC protocol may also satisfy more than one metric to provide efficient QoS.

Contribution of MAC Layer toward QoS

There are various methods by which the MAC layer can contribute to QoS directly or indirectly, and this section of the chapter discusses properties of these mechanism and how they contribute to QoS.

Service differentiation: One of the best techniques or methods used for providing QoS is service differentiation (Bhatnagar et al., 2001). By means of service differentiation within a wired or wireless network, the user requirement can be achieved. This method prioritizes and differentiates different traffic by means of classification of traffic. These different classes are treated in different manners based on their degree of importance and on the requirement for resource allocation within different classes. Service differentiation consists of two phases as follows:

1. Priority assignment
2. Differentiation between different priority levels

Priority assignment is used to assign priority among different traffic of the network based on their importance within the network. Priorities can be assigned in three different ways:

1. Static priority assignment
2. Dynamic priority assignment
3. Hybrid priority assignment

In the case of static priority assignment, priority is assigned once the packet is created and the priority does not change until the packet reaches its destination. There can be different parameters for static priority assignment:

Traffic class: Traffic class prioritizes the traffic based on the class of traffic, as in this case traffic can be of different types like real-time traffic, non-real time or best effort traffic, and so these traffic patterns are classified into different classes based on the demands of traffic within the network (Liu et al., 2006; Saxena et al., 2008).

Static priority: This assignment can be made by the type of source which generates the sensed data from sources. In this type of assignment, a particular source node or sink node can be assigned a priority based on the requirement of that node or a priority can be assigned based on the distance of that node from the sink. Packets generated from the node will inherit the properties of its creator. Data can be prioritized based on the event-based model discussed earlier in this chapter.

In *dynamic priority assignment*, the priority of the node is assigned on a real-time basis, and the priorities can be changed based on the requirements of nodes and conditions within the network. The different criteria for assigning priorities within dynamic priority assignments are as follows:

Remaining hop count: This is based on older age being given priority. In this type of priority assignment, the packets which are far away from the destination in terms of hop count, where hops refer to number of nodes, are given more priority than packets which are nearer to the sink or destination node. This is based on the fact that packets which are further from the sink or destination node are more prone to be dropped due to deadline miss or timer duration miss.

Traversed hop count: This factor is used to prioritize the traffic based on the facts that packets which traverse a higher hop count should be given priority over the packets which traverse a lower hop count in order to reach the sink node. Packets which traverse higher hop counts are more prone to drop or miss deadline. Thus, giving priority to such packets enhances the QoS by optimizing network resources, channel utilization, and increasing network lifetime and delivery ratio (Nguyen et al., 2006; Liu et al., 2005).

Traffic load: Traffic load can also be used to prioritize traffic. There are different types of nodes within a WSN: leaf node, relay node, and cluster node. Giving priority to the sensor node that is has a heavier forwarding load can reduce the rate of packet drop within the network. This is because heavier forwarding loads will overflow the buffer of the receiving node and cause packet loss. Thus prioritizing such node traffic will enhance the QoS by reducing packet loss.

Hybrid priority assignment: This method uses a combination of static and dynamic priority assignment as per the requirement of the network service delivery. This is much better than both, as this assignment is mostly needed with the current requirement of WSN to provide efficient QoS.

Differentiation method: Differentiation method is used for successful sharing of resources within the WSN as per the requirements of service delivery. This stage is the succeeding step after assigning priority in the network. Resource sharing can be done using different traffic classes at the MAC layer which can be categorized as follows:

Changing contention window size: This technique is used for setting priority within the network where there are multiple sending nodes which are sharing a single communication channel for sending their data concurrently. In order to avoid any collision among different nodes, the contention window is adopted where different contender nodes request channel allocation, and one of the nodes wins contention and reserves the communication channel for sending its data. This node reserves a contention period for sending the data and after finishing the data transmission or the end of the contention period, the next node gets the chance to reserve the communication channel for sending its data. Now to extend this medium sharing among different traffic classes with high priority as per the needs and requirements of service delivery, the contention window size can be changed or reduced to prioritize and give chances to the high priority traffic classes. If a node with a high priority traffic class gets the contention window for the communication channel, then the contention window size of such node will be increased to give a longer time period to send high priority traffic. Thus, by changing contention window size, efficient QoS can be achieved. Contention window selection based on non-uniform probability distribution can also contribute to achieving efficient QoS as normal selection is on a random basis.

Changing inter-frame space (IFS) duration: Inter-frame space duration is the space or the time for which the sensor node stays quiet before contention or back-off timer. Changing the IFS duration based on different types of traffic class according to priorities assigned to them can enhance the QoS of the network (Slama et al., 2008; Saxena et al., 2008; Nguyen et al., 2006; Firoze et al., 2007).

Scheduling of transmission slot: Scheduling can be done based on prioritizing traffic and demand of traffic for efficient delivery.

There are different protocols used in the MAC layer for *error control* which can accommodate service differentiation by changing the persistency of transmission (Yoon et al., 2004) or strength of error control codes. Error resiliency of traffic belongs to a different priority of classes which can be controlled easily and thus can be used to provide QoS for traffic.

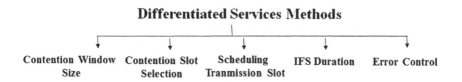

Additional factors include:

Error control: Error control is needed for providing efficient and reliable delivery of packets within the network and can be used to optimize energy dissipation. Error control can be achieved through three mechanisms:

1. ARQ technique
2. Forward error correction (FEC)
3. Hybrid ARQ technique

ARQ technique can use a stop and wait or sliding window protocol for error control which can control errors by means of acknowledgement packets. The acknowledgment tells the sender of successful transmission of a packet. The use of either protocol will depend on the type of usage.

Forwarding error control (FEC) uses different techniques for controlling the errors within the network. This technique uses redundant bits which are added to the sending data to make sending data more secure. These redundant bits are checked at the receiving side, and after confirmation are removed to fetch the actual data at the receiving node.

Hybrid ARQ uses ARQ and FEC as an error control mechanism. The sender uses a simple FEC code to encode the packet, and if the receiver gets an error message, then it sends a negative acknowledgement, and the sender uses a stronger FEC code to convert the message into an encoded form. This process continues until the receiver gets a correct packet.

Clustering: The clustering approach is used for large deployment of nodes within the network. Clustering groups nodes and provides synchronization and coordination among them. Clustering improves inter node connectivity and also facilitates data aggregation, optimizing energy dissipation in a WSN. Clustering uses static and dynamic algorithms for synchronization and coordination among nodes.

A static algorithm allocates a cluster head and members of the cluster formation for the duration of the life of the cluster. The dynamic algorithm reconstructs the cluster dynamically as per requirements and rotates the cluster head according to topological changes, thus distributing the forwarding load evenly among cluster members. This protocol consumes less energy than the static protocol.

Data suppression and aggregation: This method is used to save energy by reducing the data load of the network (Krishnamachari et al., 2002). This method suppress the amount of data by aggregating redundant data or messages belonging to the same event before transmission. It provides QoS by preventing collision and increasing resource utilization of the bandwidth.

Power control: Power control adjusts the transmission power to the minimum required for successful transmission (Muqattash et al., 2004).

Application Specific Protocol Factors

Data Suppresion & Aggregation	**Error Control**	**Clustering**	**Power Control**

QoS-Aware MAC Protocol for WSN

For provision of QoS within a WSN, the MAC layer plays a crucial role. The MAC layer supports a number of protocol for communication but not all of the protocols support QoS. One of the main issues with a WSN is energy dissipation at the nodes. Most WSN protocols are focused on energy optimization, but with the wider use of WSNs in different fields, the need for QoS has also become greater.

This section discusses different QoS-aware protocols that provide QoS either directly or indirectly.

QoS-aware protocol can be classified into 2 groups as follows:

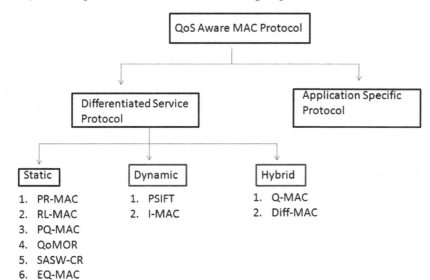

PSIFT protocol: PSIFT (Nguyen et al., 2006) is a CSMA-based MAC protocol used in event driven applications. This protocol uses a spatial correlation property of the WSN. There are two versions for suppression of unnecessary redundant packets: explicit ACK and implicit ACK using the broadcasting nature of the wireless medium. The PSIFT protocol differentiates traffic by means of varying the inter frame space and size of the contention window for different traffic classes. Prioritization is done dynamically based on the number of hop counts for a packet, the greater the number of hop counts, the higher the packet priority. Thus packets with a longer duration time to reach the destination will be given priority over other packets.

PR-MAC protocol: This protocol gives priority to different events as monitored by sensor nodes, and service differentiation is assigned to these events by means of varying contention window size and inter frame space. To assign priority among events there is an acknowledgement mechanism from the sink to each and every other node, and this acknowledgement to each and every node has the effect of wasting network resources. Sending an acknowledgement from the sink to other nodes requires powerful sink nodes that use a lot of energy, which is one of the major challenges of this protocol.

RL-MAC: This protocol is another QoS protocol that uses a CSMA scheme for packet transmission. This protocol changes the priority of nodes based on real-time observation of all nodes within the network. This protocol sends the information by adding a field to the packet header that carries information to the receiving nodes about the number of failed transmission attempts by the sender. The information provided in the field enables energy savings by minimizing the number of missed packets due to premature sleep state. QoS is assured by dividing the traffic load into different traffic categories and service differentiation by varying contention window size for each traffic categories.

Q-MAC: This protocol assigns priority according to both application layer perspective and MAC layer perspective, where the application layer deals with the content of a packet for determining priority, and the MAC layer decides priority based on traversed hop counts. Based on parameters, five queues are created including a queue that is instantly served. For intra node scheduling, the MAX-MIN fairness algorithm (Bertsekasis, 1987) is used for controlling rates, and the packetized, generalized processor sharing algorithm (Parekh et al., 1993) is used for selecting next transmitted packets. Intra node scheduling criteria is decided by the loosely prioritized random access (LPRA) algorithm which works based on transmission urgency for sending the packets. Selection of the nodes for the different queues is based on four factors: packet criticality from the application point of view, traversed hop count of packets, remaining energy of the sensor nodes, and queues' proportional load. Thus, this protocol assigns priority to the packets based on real-time application.

PQ-MAC: This protocol uses a contention- and scheduling-based algorithm for finding the best route within the network. The setup phase of this algorithm involves neighbor discovery, global clock synchronization, and assignment of slot phases followed by the transmission phase.

For assigning the slot during setup phase, it uses the DRAND (Rhee et al., 2004) algorithm which allocates different time slots based on a two hop neighbor distance node. Frame size determination is done using the time frame rule of Z-MAC (Rhee et al., 2008) protocol. The frame of PQ-MAC protocol can be divided into two sub-frames based on type of work: control frame (CF) and data frame (DF). CF is used for scheduling sleep-listen for energy efficiency and synchronization between neighboring sensor nodes. This is achieved by sequencing bits that are used to determine whether a sensor node is awake or sleep. DF on the other hand is used for secure delivery of data from the source node to the sink node. This protocol is also a real-time-based protocol which is beneficial for selection of the cluster head or relay node on a real-time basis and can also change these nodes in real time.

I-MAC: This protocol is the successor to the Z-MAC protocol. It uses benefits of the CSMA and TDMA protocols for secure transmission using QoS in the MAC layer. This protocol works in two phases: set-up phase and transmission phase. Set-up phase involves neighbor discovery, slot assignment, local framing, and global synchronization. Transmission phase uses time slots for transmission of data.

Sensor nodes in the I-MAC protocol dynamically setup priority level based on parameters like traffic load, remaining energy, and distance to sink. Based on the priority level, each node is assigned a time slot by means of the scheduling algorithm DNIB (Slama et al., 2008) and sensor nodes guaranteed within that particular slot. In case the of the owner having no data to send or the slot not being owned, than a non-owner can compete for transmission within that time slot. QoS is achieved by differentiating the services among non-owners by adopting different contention window sizes for each priority level.

Diff-MAC: This protocol is a MAC layer protocol which works based on CSMA/CA (collision sense multiple access/collision avoidance) for achieving QoS. It uses the mechanisms of service differentiation and hybrid prioritization to increase channel utilization for fast and secure delivery of data within the network. This protocol is primarily used for multimedia WSN for with heterogeneous traffic patterns. The steps in this protocol follow:

First, video undergoes large data fragmentation. The fragmented data is transmitted in bursts in order to reduce the retransmission cost in case of MAC failure.

Second, in order to reduce the number of collisions and reduce packet latencies, the protocol dynamically varies contention window size as per traffic requirements thus ensuring dynamic QoS according to traffic patterns.

Third, the protocol adopts a duty cycle for sensor nodes as per requirement and urgency of the traffic class, ensuring QoS by balancing energy consumption and delay within the network.

Alternately the protocol provides QoS by prioritizing intra-node and intra-queue for fair and fast delivery of data within sensor nodes.

SASW-CR: This protocol is based on slotted aloha used to ensure QoS for ultra-wideband (UWB) sensor networks. This protocol takes traffic as the parameter and prioritizes the nodes of the network as high and low. Differentiation of services is achieved among these prioritized nodes by using disjoint contention windows. For providing fast and reliable data delivery, a cooperative retransmission technique based on overhearing is also used.

Two queues are maintained by all nodes of the network: data queue and overhearing queue. Data queue stores packets created by sensor nodes itself, whereas the overhearing queue stores overheard packets belonging to neighboring sensor nodes. Packets can be transmitted from either queue by the nodes depending upon its mode: in selfish mode, data is transmitted from the node or packet itself, whereas in non-selfish mode data is transmitted from the overhearing node or packet from neighboring node.

EQ-MAC: This protocol provides QoS by means of service differentiation and a hybrid medium access scheme. This protocol is primarily used for a cluster-based single-hop network. This protocol can be classified into two groups: classifier MAC (C-MAC) and channel access MAC (CA-MAC).

C-MAC classifies data into four priority levels according to the importance of the packet assigned by the application layer. This creates a queue called an instant queue which contains packets of high importance that need to addressed immediately. CA-MAC protocol uses sharing of medium by means of channels and consist of four phases repeated in each frame: synchronization, request, receive scheduling and receive scheduling phases. In this protocol all nodes are synchronized and send their demand for channel request to the cluster head, and cluster head scheduling broadcast messages back to the nodes. Thus, the first three phases are only used for the exchange of control messages and sharing of medium by means of CSMA/CA. The last phase of CA-MAC is used for transfer of data by means of transmission scheduling received from the cluster head. Any sensor node having no data to be sent or unable to acquire a transmission slot will go into sleep mode to save power, thus reducing energy consumption.

Application-Specific Protocols

The EQoSA (Baroudi et al., 2007) protocol is used for transmission of video and images using QoS. It modifies the fixed session size of the BMA (Li et al., 2004) protocol and uses dynamic session sizes in relation to the number of active sensor nodes and their traffic loads. During contention period all nodes send reports of data that needs to be sent. It is the duty of the cluster head to allocate slots to sensor node sand inform them by broadcasting. Thus, the EQoSA protocol supports QoS by accommodating bursty traffic and allocating slots as per requirements of the nodes.

There are many other MAC layer protocols that support QoS indirectly, and some of them are WiseMAC (Enz et al., 2004) which is used to reduce energy consumption; CA-MAC (Kim et al., 2009); TRAMA (Rajendran et al., 2003); I-EDF; S-MAC; T-MAC; and TA-MAC which aims to deliver QoS support by using sets of QoS techniques as per the capabilities of the sensor nodes.

References

I. Akyildiz, T. Melodia and K. Chowdhury, A survey on wireless multimedia sensor networks. *Computer Networks*, vol. 51, no. 4, pp. 921–960, 2007.

U. Baroudi, EQoSa: Energy and QoS aware MAC for wireless sensor networks. In: *9th International Symposium on Signal Processing and Its Applications, ISSPA*, 2007, pp. 1–4.

D. Bertsekas and R. Gallager, *Data Networks*. Prentice Hall, 1987.

S. Bhatnagar, B. Deb and B. Nath, Service differentiation in sensor networks. In: *Proceedings of Wireless Personal Multimedia Communications*, 2001.

J. Braden, T. Leiner, F. Wilhour, R.L. Wilhour. A financial analysis of an integrated ethanol-livestock production enterprise. *North Central Journal of Agricultural Economics*. vol. 6, no. 1, 1984.

R. Braden, D. Clark and S. Shenker, Integrated services in the internet architecture – An overview. IETF RFC 1663, 1994.

P. Corke, P. Valencia, P. Sikka, T. Wark and L. Overs, Long-duration solar-powered wireless sensor networks. In: *EmNets '07: Proceedings of the 4th Workshop on Embedded Networked Sensors*, ACM, 2007, pp. 33–37.

A. Dumka, Innovation in software defined network and network function virtualization, IGI global, 2013.

A. Dumka. IoT-based traffic management tool with hadoop-based management scheme for efficient traffic management. *International Journal of Knowledge Engineering and Data Mining*, 2018.

C. C. Enz, A. El-Hoiydi, J.-D. Decotignie and V. Peiris, WiseNET: An ultralow-power wireless sensor network solution. *IEEE Computer*, vol. 37, no. 8, pp. 62–70, 2004.

A. Firoze, L. Ju and L. Kwong, PR-MAC a priority reservation MAC protocol for wireless sensor networks. In: *Proceedings of ICEE '07: International Conference on Electrical Engineering*, 2007, pp. 1–6.

Y. Gu, H. Liu and B. Zhao, Target coverage with QoS requirements in wireless sensor networks. In: *Proceedings of the 2007 International Conference on Intelligent Pervasive Computing, IPC*, 2007, pp. 35–38.

Y. Gu, Z. Yu, Q.C. Zhao, D.L. Sun, and S. Wenqin. A genetic linkage map based on AFLP and NBS markers in cauliflower (Brassica oleracea var. botrytis). *Botanical Studies*. 49. 93–99, 2008.

V. Gungor and G. Hancke, Industrial wireless sensor networks: challenges, design principles, and technical approaches. *IEEE Transactions on Industrial Electronics*, vol. 56, no. 10, pp. 4258–4265, 2009.

T. He, P. Vicaire, T. Yan, L. Luo, L. Gu, G. Zhou, R. Stoleru, Q. Cao, J. A. Stankovic and T. Abdelzaher, Achieving real-time target tracking using wireless sensor networks. In: *Proceedings of the 12th IEEE Real-Time and Embedded Technology and Applications Symposium, IEEE Computer Society*, 2006, pp. 37–48.

C. Intanagonwiwat, R. Govindan, D. Estrin, J. Heidemann and F. Silva, Directed diffusion for wireless sensor networking. *IEEE/ACM Transactions on Networks*, vol. 11, no. 1, pp. 2–16, 2003.

K. T. Kim, W. J. Choi, M. J. Whang and H. Y. Youn, CA-MAC: Context adaptive MAC protocol for wireless sensor networks. In: *IEEE International Conference on Computational Science and Engineering*, vol. 1, IEEE Computer Society, 2009, pp. 344–349.

L. Krishnamachari, D. Estrin, Deborah, S. Wicker. Impact of data aggregation in Wireless Sensor Networks. In: *Proceedings of the 22nd International Conference on Distributed Computing Systems*, 575–578, 2002.

M. Krunz, A. Muqattash and Sung-Ju Lee, Transmission power control in wireless ad hoc networks: Challenges, solutions and open issues. *IEEE Network,* vol. 18, no. 5, pp. 8–14, 2004.

J. Li and G. Y. Lazarou, A bit-map-assisted energy-efficient MAC scheme for wireless sensor networks. In: *Proceedings of the 3rd International Symposium on Information Processing in Sensor Networks*, 2004, pp. 55–60.

Z. Liu and I. Elhanany, RL-MAC: A QoS-aware reinforcement learning based MAC protocol for wireless sensor networks. In: *Proceedings of the 2006 IEEE International Conference on Networking, Sensing and Control, ICNSC'06*, 2006, pp. 768–773.

Y. Liu, I. Elhanany and H. Qi, An energy-efficient QoS-aware media access control protocol for wireless sensor networks. In: *IEEE International Conference on Mobile Adhoc and Sensor Systems Conference*, 2005.

J. Ma, C. Qian, Q. Zhang and L. Ni, Opportunistic transmission based QoS topology control in wireless sensor networks. In: *5th IEEE International Conference on Mobile Ad Hoc and Sensor Systems*, 2008, pp. 422–427.

A. Mainwaring, D. Culler, J. Polastre, R. Szewczyk and J. Anderson, Wireless sensor networks for habitat monitoring. In: *Proceedings of the 1st ACM International Workshop on Wireless Sensor Networks and Applications*, 2002, pp. 88–97.

T. Melodia, M. Vuran and D. Pompili, The state of the art in cross-layer design for wireless sensor networks. In: M. Cesana and L. Fratta (Eds.), *Wireless Systems and Network Architectures in Next Generation, Internet*, vol. 3883. Springer, Heidelberg, Berlin, 2006, pp. 78–92.

K. Nguyen, T. Nguyen, C. K. Chaing and M. Motani, A prioritized MAC protocol for multihop, event-driven wireless sensor networks. In: *First International Conference on Communications and Electronics, ICCE'06*, 2006, pp. 47–52.

A. K. Parekh and R. G. Gallager, A generalized processor sharing approach to flow control in integrated services networks: the single-node case. *IEEE/ACM Transactions on Networking*, vol. 1, no. 3, pp. 344–357, 1993.

V. Rajendran, K. Obraczka and J. J. Garcia-Luna-Aceves, Energy-efficient collision-free medium access control for wireless sensor networks. In: *SenSys'03: Proceedings of the 1st International Conference on Embedded Networked Sensor Systems*, 2003, pp. 181–192.

I. Rhee, A. Warrier, M. Aia, J. Min and M. L. Sichitiu, Z-MAC: A hybrid MAC for wireless sensor networks. *IEEE/ACM Transactions on Networking*, vol. 16, no. 3, pp. 511–524, 2008.

I. Rhee, A. C. Warrier and L. Xu, Randomized dining philosophers to TDMA scheduling in wireless sensor networks, Technical Report. Computer Science Department, North Carolina State University, Raleigh, NC, 2004.

N. Saxena, A. Roy and J. Shin, Dynamic duty cycle and adaptive contention window based QoS-MAC protocol for wireless multimedia sensor networks. *Computer Networks*, vol. 52, no. 13, pp. 2532–2542, 2008.

I. Slama, B. Jouaber and D. Zeghlache, Optimal power management scheme for heterogeneous wireless sensor networks: Lifetime maximization under QoS and energy constraints. In: *International Conference on Networking and Services (ICNS'07)*, 2007, p. 69.

I. Slama, B. Shrestha, B. Jouaber and D. Zeghlache, A hybrid MAC with prioritization for wireless sensor networks. In: *33rd IEEE Conference on Local Computer Networks*, 2008b, pp. 274–281.

I. Slama, B. Shrestha, B. Jouaber, D. Zeghlache and T. Erke, DNIB: Distributed neighborhood information based TDMA scheduling for wireless sensor networks. In: *IEEE 68th Vehicular Technology Conference*, 2008a, pp. 1–5.

S. Soro and W. Heinzelman, A survey of visual sensor networks. *Advances in Multimedia*, vol. 2009, pp. 1–22, 2009.

J. Taneja, J. Jeong and D. Culler, Design, modeling, and capacity planning for micro-solar power sensor networks. In: *IPSN'08: Proceedings of the 7th International Conference on Information Processing in Sensor Networks*, 2008, pp. 407–418.

O. Tsigkas and F. N. Pavudou, Providing QoS support at the distributed wireless MAC layer: A comprehensive study. *IEEE Wireless Communications*, vol. 15, no. 1, pp. 22–31, 2008.

T. Wark, P. Corke, P. Sikka, L. Klingbeil, Y. Guo, C. Crossman, P. Valencia, D. Swain and G. Bishop-Hurley, Transforming agriculture through pervasive wireless sensor networks. *IEEE Pervasive Computing*, vol. 6, no. 2, pp. 50–57, 2007.

A. Wood, J. Stankovic, G. Virone, L. Selavo, Z. He, Q. Cao, T. Doan, Y. Wu, L. Fang and R. Stoleru, Context-aware wireless sensor networks for assisted living and residential monitoring. *IEEE Network*, vol. 22, no. 4, pp. 26–33, 2008.

Y. Xiao, H. Chen, K. Wu, C. Liu and B. Sun, Maximizing network lifetime under QoS constraints in wireless sensor networks. In: *Global Telecommunications Conference*, 2006, pp. 1–5.

S. Yoon, C. Qiao, R.S. Sudhaakar, J. Li and T. Talty, QoMOR: A QoS-aware MAC protocol using optimal retransmission for wireless intravehicular sensor networks. In: *Mobile Networking for Vehicular Environments*, 2007, pp. 121–126.

S. Yoon, Seungju, M. Rodgers, J. Pearson, and R. Guensler. Engine and weight characteristics of heavy heavy-duty diesel vehicles and improved on-road mobile source emissions inventories: engine model year and horsepower and vehicle weight. *Transportation Research Record*. 1880. 99–100, 2004.

M. Younis, K. Akkaya, M. Eltoweissy and A. Wadaa, On handling QoS traffic in wireless sensor networks. In: *Proceedings of the 37th Annual Hawaii International Conference on System Sciences (HICSS'04)*, 2004, p. 90292.1.

N. Zogović, G. Dimić and D. Bajic, Wireless sensor networks: QoS provisioning at MAC and physical layers. In: *18th Telecommunications Forum TELFOR*, 2010.

Chapter 4

Hybrid MAC for Emergency Response Wireless Sensor Networks

Introduction

The MAC layer is layer 2 of the open system interconnection (OSI) model, and the OSI is the sublayer of the data link layer (layer 2). The MAC sublayer uses a channel access control mechanism for setting up communication among multiple nodes through a shared medium. For short-range communication like WSN, WSAN, and WBAN, the MAC protocol often uses time division multiple access (TDMA) or carrier sense multiple access/collision avoidance (CSMA/CA) for accessing of shared medium. Due to high power consumption and hardware complexity, techniques like frequency division multiple access (FDMA) and code division multiple access (CDMA) techniques are not suitable for the above mentioned short-range communication. TDMA also consumes more energy when providing synchronization. Design of MAC protocols are based on the type of application and its usage (Table 4.1).

In the case of wireless sensor networks, the nodes are randomly arranged over the area to take data from the environmental variables. The nodes used in these networks are high-powered sensors capable of wireless communication. These nodes are arranged in an ad hoc manner with no planning. Hence, the protocol used for these networks should be extremely adaptable and scalable because of constant changes in network topology, being aware of the high energy demands of sensor nodes.

Table 4.1 Comparison of Different Features

Feature	CSMA/CA	TDMA
Traffic Support	Low	High
Power Consumption	High	Low
Bandwidth Utilization	Low	Maximum
Synchronization	N/A	Mandatory needed
Mobility (Dynamic)	Good	Weak

MAC protocols are designed to handle emergency applications like disaster and fire. During an emergency response, the packet delivery ratio should be very high with low latency, and these packets should be treated with high importance. While designing protocols, these situations should be considered, and these packets should be treated with the more importance than any other data on the network. Fairness is also a very important factor, as MAC allows access to all information from all the sensor nodes that sensed important data and sends the consolidated data to the sink node.

Designing a MAC protocol for such highly deliberated events must take the following important factors into account:

1. *Traffic load*: During the emergency event, traffic load on the network may increase significantly and may become unbalanced. The protocol should be designed so that it can handle an unbalanced and increased load without performance of the network becoming affected.
2. *Energy efficiency*: Energy efficiency is one of the most important factors in a WSN. Under normal operations the energy consumption of each node is limited, but in the case of emergency situations the energy efficiency of each node decreases considerably, as each node has to support increased and unbalanced data with high delivery ratio and low latency.
3. *Delay latency*: Delay latency occurs when between the sensor node and the sink node. In the case of an emergency, latency in delivering information from sensor node to sink node should be minimal, as high priority packets need to timely delivery to the sink node.
4. *Delivery ratio*: Delivery ratio is the ratio of information being sent from the sensor node to the sink node, and this ratio should be maximized for normal and emergency situations.
5. *Detection delay*: Any event detected should be reach the sink node from the sensor node at a predictable duration. Detection delay in emergency situations should be minimal.

Based on the above factors and the adaptation of traffic and topology, various MAC protocols had been designed. Based on different criterias and technologies used, several researchers had proposed several MAC protocols depending upon the design requirements of the WSN. These protocols depends upon how the nodes are access the channels. WSN MAC protocols fall into four categories as follows (Figure 4.1):

1. Contention-based
2. Channel-polling-based
3. Schedule-based
4. Hybrid

Contention based MAC protocols: The contention-based MAC protocol is a carrier sense multiple access (CSMA) protocol. The contention-based MAC protocol can easily adapt to changes in node density. These protocols can easily cope with an increase in traffic load.

Nodes in contention-based MAC protocols work on acquiring channels. Network nodes compete with their neighbors to acquire channels to send data. On sensing a channel to be idle, a node can send the data, otherwise the node will defer the transmission.. The deferment time is usually determined by a "back-off" algorithm. These protocols are dynamic and flexible for scalable networks.

Contention-based protocol does not require clustering or topology information, and each node can decide independently for contention without controlling frame exchange, and transmission is solely handled by the sender.

Examples of contention-based MAC protocols are S-MAC, T-MAC, B-MAC, WiseMAC, TA-MAC, X-MAC, and MaxMAC.

Channel-polling-based MAC protocols: This uses preamble sampling and low power listening (LPL) for its operations. Preamble means the node sends prefixed data packets

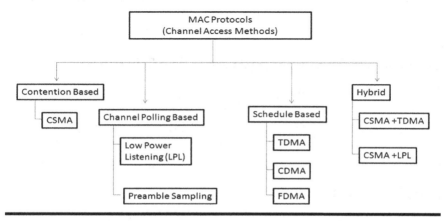

Figure 4.1   Classification of MAC protocol.

with an extra byte. This preamble byte is sent over a channel to the destination node in view of detecting radio activity and a wake-up call to the destination before the destination receives the actual payload from the sender. On receiving the preamble, the receiver activates its radio for receiving the packets from the sender; otherwise, the receiver goes into sleep state until the next polling occurs. There is no need for clustering scheduling and synchronization in this protocol, as active/sleep does not occur.

An example of a channel-polling-based MAC protocol is the Berkley MAC (B-MAC) protocol.

Schedule-based MAC protocols: Schedule-based MAC protocols can make use of time division multiplexing (TDM), frequency division multiplexing (FDM), or code division multiple access (CDMA). Due to complexity of FDMA and CDMA, it mostly uses TDMA, which schedules each channel for the node to send, receive, or sleep in order to save energy. Some protocols in schedule-based MAC protocols are adaptable to changes in topology, whereas some are not. A schedule-based MAC protocol is unable to maintain a schedule if there is an increase in traffic in the case of emergency information dissipation.

Schedule-based protocol is useful in providing a network with minimum collision, implicit avoidance of idle listening, and less overhearing problems inherent in WSNs. On the other hand it provides a bounded and predictable end-to-end delay. This protocol increases the average queuing delay, as a node has to wait for its allocated time slot before accessing the channel, and this may lead to low throughput, extra and overhead traffic, lack of adaptability, and reduced scalability.

Examples of schedule-based MAC protocols are TRAMA, FLAMA, and VTS.

Hybrid MAC protocols: A hybrid MAC protocol is a joint improvement protocol that combines the strength of two or more MAC schemes. Generally, a hybrid MAC protocol combines a synchronous scheme with an asynchronous scheme to achieve high throughput in the network.

Examples of hybrid MAC protocols are Z-MAC, P-MAC, Funneling MAC, Crankshaft, RRMAC, EB-MAC, BurstMAC, i-MAC, and ER-MAC.

Study of Various Existing Hybrid MAC Protocols

The MAC protocol being used in a WSN is important, as the selection of a MAC protocol is decided based on the criteria for energy conservation and increased efficiency for the WSN. An efficient MAC protocol may be used for providing flexibility, reducing collision, or increasing the throughput for various sorts of applications being used on the WSN.

Z-MAC (Zebra MAC)

The Z-MAC protocol uses the strengths of the TDMA and CDMA protocols for providing efficiency in the WSN. For low contention in the network, Z-MAC uses CSMA features, whereas for high contention traffic it uses the TDMA protocol.

Z-MAC protocol is dynamically adapting to topology changes and time synchronization failure within WSNs.

The CSMA protocol i used by Z-MAC as a baseline MAC scheme, whereas TDMA is used to enhance contention resolution. The Z-MAC protocol assigns a time slot at the time of deployment to a particular node, thus a higher overhead is incurred at the time of deployment. For providing efficient channel scheduling, Z-MAC uses a DRAND algorithm which is a centralized, channel reuse, scheduling algorithm. The node assigned a slot can use the assigned slot/channel periodically after every predetermined period, and this is called a frame. A single channel can be used by multiple nodes, but the node assigned to the time slot is the "owner" of the slot, whereas other nodes are non-owners. The DRAND algorithm allows more than one owner of a slot. The Z-MAC protocol allows a node to transmit at any time slot, but before sending the node performs carrier sensing and sends only when the channel is clear. The priority of the node is implemented by adjusting the initial contention window size such that the owner of the channel is always preferred over a non-owner. In the case of the owner having nothing to send, then a non-owner can send a packet through the channel. This priority scheme has the effect of switching between CSMA and TDMA depending upon the level of contention. In a case where we want to change the priority of owner and non-owner, then that can be done independently. Thus, the Z-MAC protocol increases the robustness of the protocol in synchronization and reduces slot assignment failure while increasing scalability of contention (Rhee et al., 2008).

During initial setup of the Z-MAC protocol, various operations run in sequence as neighbor discovery, slot assignment, local frame exchange, and global time synchronization. All of these operations run during setup phase and will not execute other than if the network topology changes.

In neighbor discovery, a simple neighbor discovery protocol runs that periodically broadcasts a ping to its one hop neighbor in order to fetch its one hop neighbor list. The DRAND protocol is used to assign a slot to a frame, and this is done based on the local neighborhood size of each node. The DRAND algorithm can also perform localized time slot assignment without modifying a time slot already assigned to a pervious existing node in case of the addition of new nodes. Once the slot is assigned to each node, the nodes will decide on time slots for transmission— "time frame"— for exchanging of frame and information. In order to synchronize, all frames should keep the same time frame so that their time slot 0 is always at the same time.

Thus, the Z-MAC protocol proposed for the wireless sensor network is able to dynamically adjust behavior of the MAC protocol between CSMA and TDMA, based on the level of contention within the network. The Z-MAC protocol increases the performance of the MAC protocol under high contention by using its knowledge of topology and loosely synchronized clocks. Whereas under low contention, Z-MAC behaves like CSMA, as there is no need for any special attention. Thus, this protocol provides better dynamic efficiency in low as well as high contention.

P-MAC (Pattern MAC)

P-MAC is a hybrid wireless sensor protocol that uses a time-slotted technique based on traffic and traffic pattern of its neighbors and can adaptively determine a sleep/wake-up schedule for a node in order to achieve higher throughput at high traffic and minimize energy consumption during lighter load traffic patterns.

The schedule for each node to be active depends upon the node's traffic and its neighbor traffic pattern. The P MAC protocol works on the concept of a "pattern exchange" framework for increasing the efficiency of the MAC protocol in wireless sensor networks.

The P-MAC protocol is useful as it is adaptable to traffic patterns enabling a sensor node to wake up quickly as and when the traffic load on the network increases, thus making the network adaptable to variable traffic patterns. The P-MAC protocol saves power by means of localization by waking up sensor nodes frequently which are involved in communication, while allowing sensor nodes to sleep which are not involved in data gathering or relaying process for a longer duration. The P-MAC protocol also allows power savings by means of reducing idle listening. This is done by allowing sensor nodes which are not involved in communication to sleep for a longer duration, thus reducing energy wasted due to idle listening during periodic wake up. The P-MAC protocol provide time synchronization among sensor nodes.

Funneling MAC

Funneling-MAC is a hybrid protocol for sensor networks that is capable of mitigating the funneling effect and boosting application fidelity in the sensor network. This protocol uses CSMA and TDMA techniques for providing better efficiency. This protocol is based on the carrier sense multiple access/collision avoidance (CSMA/CA) pattern implemented network-wide, whereas the TDMA algorithm is overlaid on a small number of hops from the sink (funneling). Funneling MAC protocol is a hybrid MAC protocol which is free from scalability problems and is a sink-oriented protocol. Funneling MAC is a localized protocol, as TDMA operates locally within a funneling region close to the sink and not across the entire sensor field.

Wireless sensor networks operate in a funneling manner where any event generated by any sensor node travels hop-by-hop in a many-to-one traffic pattern towards a centralized sink node. This hop-by-hop communication with centralized collection of data at sink node creates a choke point on the free flow of events out of the sensor network that may result in collision, congestion, and loss of packets within the network. The sensors that are near to the sink have large data or packets that need to be forwarded compared to nodes further away from the sink, and this region of funneling is called an intensity region due to the large amount of data. Due to heavy traffic in this region, sensor nodes consume more energy than nodes further away from the sink node which may result in shortening the lifetime of the sensor network. The funneling MAC protocol focuses on this particular problem of wireless sensor networks to increase the efficiency and lifetime of a sensor network.

The funneling MAC protocol finds the where the funneling effect occurs within the sensor network. The funneling MAC protocol works within the intensity region of the funneling event. Pure CSMA/CA operates within the entire sensor network along with acting as a component of the funneling MAC working within the intensity region. The funneling MAC protocol uses local TDMA scheduling within the intensity region only, thus providing additional scheduling opportunity to the nodes closer to the sink node where intensity is greater. Implementation of the TDMA protocol within local high intensity regions of the entire network provides a scalable solution for deployment of TDMA scheduling in a sensor network, thus helping to boost application fidelity as measured at the sink (Ahn et al., 2006).

The funneling MAC protocol is a hybrid protocol that uses the CSMA and TDMA schemes in the intensity region under the control of sink within the sensor network to increase the throughput and loss performance of sensor networks, even under light traffic conditions and for small intensity regions with a depth of one or two hops. Along with this, the funneling MAC protocol is capable of managing the dynamic variation in the intensity region of a sensor network.

Crankshaft

Crankshaft is a hybrid routing protocol that focus on the inefficiency of a dense network in order to improve the efficiency of a dense network employing node synchronization and an offset wake-up schedule to overcome the overhearing problems of neighboring nodes. This protocol also saves energy by means of using efficient channel polling and contention resolution techniques. The crankshaft protocol reduces overhearing and communication grouping by letting nodes power down their radio signals alternatively rather than simultaneously. The protocol is named "crankshaft" as it works on the principle of the internal combustion engine where the moment a piston fires is a fixed offset from the start of rotation of the crankshaft.

The crankshaft protocol works on the concept that nodes are awake only to receive message at fixed offsets from the start of the time frame. Crankshaft divides time into frames and each frame into slots. The slots are divided into broadcast and unicast. In a broadcast slot, every node will wake up to listen for incoming messages, and the node that contains a broadcast message contends with all the other nodes to send that message. A frame starts with all the uncast slots, followed by the broadcast slots.

Each node also listens for one unicast slot every frame. Within this slot, if a neighboring node wins the contention, then neighboring nodes can send message to that node. The crankshaft protocol uses a data/ack sequence for sending messages in unicast messaging, and the length of the slot should be longer for the contention period, maximum-length data message, and acknowledgement message. On the other hand, the sink node should listen to all the unicast message as the sink node is the destination for nearly all of the traffic and hence requires more bandwidth to handle heavier traffic (Halkes and Langendoen, 2007).

Thus, the crankshaft protocol provides good convergecast performance at low energy consumption in dense networks by minimizing broadcast flooding. Crankshaft protocol also employs slot scheduling for reducing overhearing of sensor nodes. Crankshaft protocol also reduces energy consumption by reducing overhearing of sensor nodes. It also provides good broadcast flood delivery. It can be concluded that the crankshaft algorithm is more suitable for long-lived monitoring applications, where energy efficiency plays a vital role as energy consumption in this protocol is lower. It also provides a required delivery ratio for convergent traffic at extremely low energy consumption.

RRMAC

RRMAC is a TDMA- and CSMA/CA-based hybrid protocol that works efficiently for a multihop convergecast network. It guarantees end-to-end packet transmission delay by avoiding packet collision. This protocol assigns sequences to the time slots to maintain a continuous flow of packets from leaf node to top node. In the case of convergent networks, assignment of sequence allows top level nodes to acquire all data in one superframe duration. In order to reuse the time slot with nodes having two different communication ranges within a sensor network, RRMAC protocol delays the acknowledgement which enables the reduction of superframe duration and end-to-end delay (Kim et al., 2008).

RRMAC protocol uses TDMA for avoiding collision with a network which can be achieved by exclusively assigned time slots for transmitting devices that try to transmit latency sensitive packets. Thus, RRMAC can be used in real time systems and provide better delivery success rates and power consumption.

EB-MAC (Event-Based Media Access Control)

This protocol is based on event-based systems that suffer from problems associated with long periods of inactivity and short abrupt periods of high data contention when an event is detected. As sensors collecting data send high data packets as and when they detect an occurrence of an event. EB-MAC protocol uses an election-based scheduling technique for arranging the data transfer dynamically. This protocol is implemented using a MicaZ sensor. This protocol shows high performance, throughput, reliability, and energy conservation than previously existing protocols.

EB-MAC protocol is tailored for event-based systems with a view that it can handle high contention data along with maintaining a good performance in low traffic conditions. As an event is detected by a node, a round is initiated by the node through which a schedule is created for nodes wishing to transmit. Each of these nodes have their own slots to transmit without any contention, and the purpose of this protocol is to provide a contention-free channel for transmission of contention-free data, reducing energy waste and increasing latency. EB-MAC protocol uses B-MAC (Polastre et al., 2004) CCA, and LPL techniques for accessing channel activity and to duty cycle the radio.

In the case of multi-hop configuration, there are some implications of the EB-MAC protocol as any node that belongs to two different clusters may have different schedules at the same time which may result in transference of data at the same time in the node. One solution to this problem can be to let each cluster send data one at a time at cluster level in a manner so that the neighboring cluster is prohibited from sending data at the same time in order to maintain synchronization. This is not an optimal solution and may result in latencies that may degrade the performance of the system. Another solution may be that nodes belonging to two different clusters can adopt two schedules, one for each cluster and thus synchronization will be maintained by sending data using the group ID which is their cluster ID (Merhi et al., 2009).

Burst MAC

Burst MAC protocol is a hybrid protocol that makes use of CSMA- and TDMA-based techniques for increasing the throughput and lower energy overhead for event-triggered applications which are have correlated traffic bursts. In order to handle correlated burst traffic, burst the MAC protocol makes use of the TDMA technique and multiple radio channels. This protocol is best suited for applications where large amounts of data need to be extracted or transferred from sensor nodes. Thus, for large amounts of data transmission this protocol is suited to provide the best possible solutions. The burst MAC protocol can work with fixed as well as dynamic topology.

The burst MAC protocol makes use of multiple techniques grouped together for providing high throughput and efficiency for a wireless network. The burst MAC protocol makes use of multiple radio channels, cooperative transmissions and techniques, scheduling of time slots, and assignment of channels to integrated together to provide efficient throughput for a wireless network. Burst MAC uses synchronous rounds to prevent any collision within the network (Ringwald and Romer, 2009).

i-MAC

i-MAC is a hybrid wireless MAC protocol that learns to expect and plan for traffic bursts and consequently coordinates node transmission efficiently. This protocol is designed with manufacturing systems in mind, as manufacturing systems are repetitive in nature in terms of their operations and hence the communication traffic patterns are repetitive in nature, so this lies at the heart of this protocol. A MAC protocol needs to learn this repetitive pattern and thus schedule transmission accordingly and efficiently. Thus, i-MAC protocol work with the same pattern and avoid transmission collisions in a channel. i-MAC learns on the fly and continuously adapts itself to the host machine and its operating conditions (Chintalapudi, 2010).

ER-MAC

The ER-MAC protocol is a hybrid MAC protocol that is designed to handle emergency response system. The earlier version of hybrid MAC protocol focused on topology and traffic pattern changes but did not focus on packet prioritization. ER-MAC protocol design is based on packet prioritization. In order to handle large amounts of data packets, the ER-MAC protocol allows contention in TDMA slots, thus trading energy efficiency for a higher delivery ratio and lower latency rate. In order to segregate emergency event data from other data, the ER-MAC protocol segregates packets based on their priority into two queues as high priority and low priority packets. The ER-MAC protocol supports fairness so that a sink can receive complete information from all sensor nodes. For nodes to leave or join the network on a real time basis, the ER-MAC protocol supports synchronized and loose slot structure so that nodes can easily modify their schedule locally.

ER-MAC is hybrid MAC protocol that is flexible in adapting to traffic and topological change. ER-MAC functions differently in normal and emergency situations. In a normal situation, it focuses on saving energy and so nodes only wake up for their scheduled slots and otherwise sleep to save energy. While in emergency situations, nodes participating in emergency monitoring, change their MAC behavior by allowing contention in each slot to achieve low latency and a high delivery ratio (Sitanayah et al., 2010).

ER-MAC protocol is designed keeping emergency-based situation in mind and priority assignment which was left in earlier version of hybrid protocol. It prioritizes emergency traffic and offer synchronized and loose slot structure to locally modify the schedule by nodes. ER-MAC protocol achieve higher delivery ratio and lower latency at low energy consumption compared to other hybrid protocols (Table 4.2).

Table 4.2 Comparison of Different Hybrid Protocols

Hybrid Protocols	Prime Objective (Optimization)	Traffic Adaptability	Topology Adaptability	Packet Priority	Fairness
Z-MAC	Throughput	Good	Good	No	Yes
PMAC	Energy, throughput	Good	Medium	No	Yes
Funneling-MAC	Throughput	Good	Good	No	Medium
Crankshaft	Energy	Medium	Good	No	Medium
RRMAC	Delivery, latency	Medium	Good	No	Yes
EB-MAC	Delivery, latency	Medium	Good	No	No
Burst MAC	Overhead, throughput	Good	Good	No	Yes
i-MAC	Latency	Medium	Good	No	Medium
ER-MAC	Energy, throughput, latency	Good	Good	Yes	Yes

References

G.-S. Ahn, S. G. Hong, E. Miluzzo, A. T. Campbell and F. Cuomo, Funneling-MAC: A localized, sink-oriented MAC for boosting fidelity in sensor networks. In: *Proceedings of the 4th International Conference on Embedded Networked Sensor Systems*, 2006.

K. K. Chintalapudi, i-MAC: A MAC that learns. In: *Proceedings of the 9th ACM/IEEE International Conference Information Processing in Sensor Networks (IPSN'10)*, 2010, pp. 1–9.

G. P. Halkes and K. G. Langendoen, Crankshaft: An energy-efficient MAC-protocol for dense wireless sensor networks. In: *Proceedings of European Conference on Wireless Sensor Networks EWSN'2007*, 2007, pp. 228–244.

J. Kim, J. Lim, C. Pei Czar and B. Jang, RRMAC: A sensor network MAC for real time and reliable packet transmission. In: *Proceedings of International Symposium Consumer Electronics*, 2008, pp. 1–4.

Z. Merhi, M. Elgamel and M. Bayoumi, EB-MAC: An event based medium access control for wireless sensor networks. In: *Proceedings of the 2009 IEEE International Conference, Pervasive Computing and Communications (PerCom'09)*, 2009, pp. 1–6.

J. Polastre, J. Hill and D. Culler, Versatile low power media access for wireless sensor networks. In: *Proceedings of the ACM Conference on Embedded Net-worked Sensor Systems (SENSYS '04)*, November 2004, pp. 95–107.

I. Rhee, A. Warrier, M. Aia, J. Min and M. L. Sichitiu, Z-MAC: A hybrid MAC for wireless sensor networks. *IEEE/ACM Transactions on Networking*, vol. 16, no. 3, pp. 511–524, 2008.

M. Ringwald and K. Romer, Burstmac: An efficient mac protocol for correlated traffic bursts. In: *Networked Sensing Systems (INSS), Sixth International Conference on*, 2009, pp. 1–9.

L. Sitanayah, C. J. Sreenan and K. N. Brown, ER-MAC: A hybrid MAC protocol for emergency response wireless sensor networks. In: *Proceedings of the 2010 Fourth International Conference on Sensor Technologies and Applications*, July 18–25 2010, pp. 244–249.

NETWORK LAYER

Chapter 5

Routing Schemes in Wireless Sensor Networks

In order to have a secure routing protocol the following parameters must be considered:

1. Node deployment
2. Energy consumption
3. Data reporting model
4. Fault tolerance
5. Scalability
6. Data aggregation
7. Quality of service (QoS)
8. Security

Node Deployment: Nodes can be deployed in a deterministic or randomized manner depending upon the conditions and requirements of the WSN applications. In the deterministic approach, sensors are located manually and routing of data is done via a predetermined path (Wood et al., 2006; Dhurandher et al., 2010; Gaurav et al., 2012). The randomized approach is used for short-range communications where sensor nodes are densely deployed (Lou and Kwon, 2006; Al-Wakeel and Al-Swailemm, 2007; Fan et al., 2010; El-Bendary et al., 2011). The arrangement of nodes in a randomized manner is more easily deployed than in the deterministic approach. The selection of either technique for node deployment is application dependent.

Energy consumption: This is one of the important factor in terms of WSNs, as nodes used in WSNs use batteries of limited duration inoperation, so energy consumption is an important concern for WSNs. Within a sensor network, energy is needed for communication and transmission. In order to overcome problems of high energy consumption, there are two types of protocols proposed for energy conservation:

a. Light weight—works on attributed which are responsible for energy consumption and works on these attributes to save energy.
b. Efficient—focuses on alleviating the resource inadequacy problem of sensor nodes to save energy consumption.

Data reporting model: This model is a data dissemination model of routing protocols used in WSNs. This model also depends on time criticality of data along with type of applications being used in the WSN. The data reporting model can be categorized into four types:

a. Time driven: This reporting model is useful where sensor nodes need to send data periodically.
b. Event driven: This reporting model is used in case of occurrence of certain events which may cause sudden or drastic changes in the sensed attributes from sensor nodes.
c. Query driven: In this reporting model, nodes respond immediately to sudden changes in the value sensed due to the occurrence of new queries by the user for specific requirements.
d. Hybrid: This model is a combination of either two or three the models described above based on the condition and application usage.

Fault tolerance: Fault tolerance in a WSN is defined as the ability of sensors nodes to retain their functionality without interruption in situations like node failure due to physical damage, lack of power, and environmental interference. Fault tolerance can be achieved by means of redundant node deployment as an option to achieving fault tolerance. Various parameters which should be keep in mind while developing a fault tolerant system are:

a. Key structure: This is how the attributes in a protocol cooperate with keys to make a system fault tolerant.
b. Intrusion tolerant: This parameter is used to indicate the capability of nodes in a WSN to make the routing protocol operative even in case of failure of some nodes.
c. Multipath: It determines an alternative path to data in case primary fails and thus provide fault tolerance
d. Hybrid: This incorporates a combination of more than one parameter in order to achieve fault tolerance.

Scalability: Scalability supports increasing or decreasing WSNs in terms of number of nodes within the WSN. Thus the routing protocol used should be adaptable for a scalable network to support more sensor nodes within a network. Based on the number of nodes within a network, it can be classified as *limited* with less than 100 nodes, *medium* with 100 to 999 nodes, and *good* with 1000 or more nodes in the network.

Data aggregation: Data aggregation is when data is combined as it arrives from several nodes of the WSN. This can be achieved by using functions min, max, average, and suppression. Data aggregation removes the communication overhead of the routing operation. This can be achieved by means of labeling a sensor as a cluster head that accommodates all data from different sensors and transmits a single value to the base station (BS). Thus, this reduces the number of transmissions and data redundancy, and it also reduces the communication overhead for sensors and thus minimizes energy consumption. Data aggregation can be achieved using *tree-based* (Di Pietro et al., 2003) structure routing protocol or *cluster-based* (Xizheng, 2007) structure routing protocol.

Quality of service: There are various parameters that need to be addressed for achieving quality of service as reliability, latency, and bandwidth. As per the criticality of applications for WSN, there may be a requirement for time-specific data transmission without any delay. In these conditions, a certain bounded latency time is required for specifying validity of the data. For increasing the efficiency of routing operations, high bandwidth is required. Thus, there is a need to design an efficient routing technique for providing efficient QoS services within the network.

Security: Since nodes within the WSN use the wireless mode for communication, security is one of the prime concerns. The security issue can affect the performance of the network, especially for real time transmission in the network, the security issue should be of utmost importance. The WSN is prone to attack due to its features of wireless communication, dynamic topology, and its broadcasting nature. The securities objectives needs that to be kept in mind while designing the protocol are:

1. Authentication
2. Integrity
3. Confidentiality
4. Availability
5. Freshness

Authentication is needed as an advisory node can alter data from any node by injecting fabricated packets into transmitted messages within a network that can cause a threat to the packets. Thus, authentication of packets is needed for wireless communication secure transmission.

Integrity of data refers to securing data so that it is not lost or corrupted which may lead to errors in the communication process. Thus, this is a mechanism to prevent data from being altered by attackers, as any alteration will affect the entire communication mechanism.

Confidentiality is needed to ensure the data is received and accessed by authorized entities only. Thus, confidentiality prevents data content being revealed to intruders or eavesdroppers during transmission.

Availability is related to the sustainability of networks in the case of failure of nodes or attacks from intruders. Whereas freshness ensures that there will be no replay of old data from any advisory node, and all data received should be fresh and new. Freshness is achieved when the WSN adapts the shared key mechanism for communication of messages, whereas advisory send old keys which is disseminated to all the nodes.

All of the above security objectives can form guidelines for researchers to achieve security and fight attacks within a network.

Based on the attributes and parameters discussed above, various routing protocols are designed for WSNs based on the types of usages and application for which these protocols are used.

Routing Protocols in WSNs

Routing protocols within WSNs fall into two groups as shown in Figure 5.1.

The first classification of routing protocol includes flat, hierarchical, and location based, as this refers to the network structure. The parts of a routing protocol

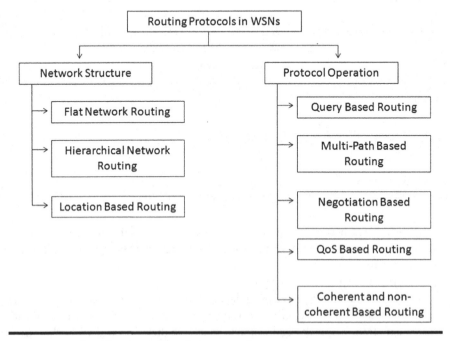

Figure 5.1   Classification of routing protocols.

are query based, multi-path based, negotiation based, QoS based, and coherent based, each of which refers to an operation.

In the flat-based routing protocol, all the nodes of a network are assigned equal loads or responsibility, whereas in hierarchical-based routing protocol all the nodes play different roles within the network depending upon its structure. Location-based routing protocol uses the location of sensor nodes for routing the data within the network.

Operation-based routing protocols are also called adaptive routing protocols, as they adapt to certain networking conditions and energy levels, and these protocols can be controlled or managed by system parameters. These protocols can be further classified into three groups as proactive, reactive, and hybrid based on how the source searches for the path to a destination. The proactive protocol takes the predefined routes from the source to the destination similarly to static routes, whereas reactive protocol finds the route based on demand. Hybrid protocol is a combination of proactive and reactive protocols where some routes are predefined and some are reactive.

Network Structure-Based Routing

Flat-Based Routing Protocol

In flat-based routing protocol all the nodes perform the same task and collaboratively perform the sensing task for the network. It uses data centric routing where all routing is confined to the center node which is the base station. The base station sends a query for specified data to a particular region and waits for the nodes of that region to respond to the query. In order to specify the data, attributed-based naming is used for the purpose. Some of the examples of flat-based routing protocols are sensor protocols for information via negotiation (SPIN), direct diffusion, rumor routing, stream-enabled routing (SER), gradient-based routing (GBR), coherent/not coherent processing, constrained anisotropic diffusion routing (CADR), COUGAR, active query forwarding in sensor network (ACQUIRE), threshold sensitive energy efficient network protocol (TEEN) and adaptive periodic TEEN (APTEEN). Discussion of the working and methodology of some of these protocols follows.

SPIN

SPIN protocol was proposed by Heinzelman et al. (2002) and is based on a flat-based routing protocol where all the nodes within the network can work as the base station. Thus, a query can be sent to any node to retrieve information immediately. SPIN protocol uses three types of messages ADV, REQ, and DATA for setting up the communication between neighboring nodes. ADV messaging is used for advertising any new data within the network. REQ is request messaging for data from sensor nodes. DATA messaging is used for sending actual data within the network using SPIN protocol in the following manner. Any neighbor wanting to send data will send a REQ request to a specified sensor node, the sensor node will send the DATA packet to the request node, and this process repeats until the complete transfer of data within the network. There are various versions of SPIN protocol that focus on different characteristics to increase efficiency and overcome the drawbacks with earlier versions of SPIN.

Direct Diffusion

Directed diffusion (Intanagonwiwat et al., 2000) uses a data aggregation paradigm for WSN. This protocol works on the concept of combining or aggregating the data coming from different sensor nodes and routes them by eliminating redundancy and reducing the number of transmissions, saving energy and increasing the lifetime of the network. This protocol is a data-centric protocol, where an application in the sensor attribute value pairs for the data generated by the node. Thus directed, diffusion routing traces a single route from multiple sources to a single destination within a network by consolidating the redundant data.

Sensor nodes within a network senses data and create information for environmental parameters. Base station requests for specified data needed from sensor nodes. Sensor nodes broadcast the needed information to all its neighbors across the network and it then reaches the base station. The base station receives information which is specified by attribute value pairs and a direction. This information can flow from different paths and from these paths, the best path will be selected in order to avoid flooding the network. Data aggregation is used for reducing the communication paths within the network.

Rumor Routing

Rumor routing is a variation of diffusion routing and is used where geographic routing is not feasible. Diffusion routing uses the concept of flooding for information dissipation within a network, but in cases of a smaller amount of data flooding it is not a good solution. In such a case, routing of queries to the node is done through observation of a particular event for retrieving information rather than the flooding approach. As the query reaches the node, the event is sent to reach the information from sensor node to the base station through a routed path.

As sensor nodes detect an event, they update that event in their event table and generate agents. Agents are packets that are used for a flooding event within a network. Thus, these agents propagate information over the entire network and find the best path. As nodes receive queries through routed paths, these nodes respond to the query by using their event table and sending the event to its destination through this routed path to avoid flooding. This protocol works well only in the case of small events, and as to large events, the cost of maintaining an event table and agents in each node is costly and hard to maintain. In order to maintain an efficient time duration for receiving information from source to destination in rumor routing, we use the concept of time to live (TTL).

Stream-Enabled Routing

SER is a routing protocol that was proposed by Su and Akyildiz (2002) that works based on the concept that the source node selects the route based on the instruction given by the sink node. The amount of energy used is one of the key parameters for selecting the route from the source to destination. Thus, the sink uses flooding in the sensor field in order to find the source. Once the source is found, the path from the source to the sink is established for transmission of data.

Gradient-Based Routing

The GBR protocol proposed by Schurgers and Srivastava (2001) works on the concept of hop counts, where each node calculates the height with the base station which is the number of hops or number of counts of sensor nodes it takes to reach to base station. This protocol uses the concept of data aggregation and traffic distribution to optimize resource utilization for maximum throughput within the network. In order to distribute traffic in a balanced manner within the network, three different schemes are used: stochastic, energy-based, or stream-based depending upon the parameters that need to be kept in mind.

Constrained Anisotropic Diffusion Routing

The CADR protocol is designed on the concept of direct diffusion based on a data-centric approach and was proposed by Chu et al. (2002). The CADR routing protocol optimizes the information gain by minimizing bandwidth utilization and latency.

Each node within CADR evaluates parameters like information or cost objectives, and route data information required for sending information within the network by means of local information or cost gradient, and end user requirements. Thus, based on the value given by neighboring sensor nodes, the CADR routing protocol takes local decisions by using a greedy search method. For forwarding the packets to neighboring nodes, two different approaches can be used with highest objective function as the first alternative, steepest or local gradient in the objective as the second alternative and maximizes the combination of local gradient of objective function with distance improvement to estimated optimum location.

COUGAR

The COUGAR protocol was proposed by Yao and Gehrke (2002). It is also a data-centric protocol that considers the sensor network as a big, distributed database system. This protocol introduces a new layer of query between the application and sensor network that make use of a declarative query for abstracting query processing from the network layer and utilizes it in the network data aggregation process in order to save energy.

In COUGAR protocol, the base station (BS) generates a query which is used to find a leader among sensor nodes for carrying information within the network. This leader is selected to perform aggregation of data and transmitting of data to the BS. The query generated by the BS can also be used for data flow and in-network computation for incoming queries and sending data to the relevant node.

Active Query Forwarding in Sensor Network

The ACQUIRE protocol was proposed by Sadagpan et al. (2003) and uses the same concept as COUGAR of a distributed database for dividing complex queries into several sub-queries for forwarding of data within the network. This is achieved by allowing several nodes to respond to queries at the same time. The performance of this protocol is better than directed diffusion as directed diffusion protocol cannot be used for complex queries due to consideration of energy consumption. In addition, the directed diffusion protocol uses a flooding mechanism for continuous aggregation of queries. The ACQUIRE protocol uses efficient querying by adjusting the values of look ahead parameter d, where d is equal to the diameter of the network. If the network diameter is small, that is if value of d is too small, then the number of hops that queries have to travel will be greater.

In ACQUIRE, the base station sends a query which is forwarded by each node to those who are receiving the query. Each node receiving the query will respond to the query partially by means of pre-fetched data it gathers and forward the data to other sensor nodes. For updating pre-fetched information, the node depends on its neighbors and gathers updated information from its neighbor by means of look ahead of d hops. After resolving the query receive from BS, it sends back it through the shortest reverse path back to the BS.

Hierarchical Network Routing

Hierarchical network routing is also called cluster-based routing and works by making clusters of nodes. These clusters will have one cluster head responsible for aggregation and reducing the data in order to save energy. This routing works in a two layer format, where one layer is responsible for selection of the cluster head while the other layer is responsible for routing. Different nodes in different hierarchies within the network perform different tasks. High energy nodes within the network can be used for processing and sending the information, whereas low-energy nodes can be used for sensing purposes in the proximity to the target. Hierarchical network routing protocols are best suited for saving energy within the network. Different routing protocols that work in hierarchical routing are simple hierarchical routing protocol (SHRP); low-energy adaptive cluster hierarchy (LEACH); power efficient gathering in sensor information systems (PEGASIS); hierarchical energy aware protocol (HEAP) for routing and aggregation in sensor networks; hierarchical periodic, event-driven and query-based (HPEQ); threshold sensitive energy efficient sensor network protocol (TEEN); small minimum energy communication network (SMECN); self-organizing protocol (SOP); and SPIN. Discussion of the working and methodology of some of important protocols of hierarchical routing follows.

Low-Energy Adaptive Cluster Hierarchy

This protocol was proposed by Heinzalman et al. (2000) and is one of the basic protocols for cluster hierarchical routing. LEACH works in a clustered manner for forwarding of information and also includes distributed cluster formation. In the LEACH protocol, there is random selection of sensor nodes for the cluster head (CH). The cluster head is evenly distributed among sensor nodes in order to save energy consumption. The CH aggregate data arriving to it and then compress the data to transmit it to the BS.

In order to reduce collision among intra- or inter-cluster networks, LEACH uses a TDMA/CDMA approach. LEACH performs better when there is no need for constant monitoring of sensor data. In LEACH all the data is collected centrally and periodically in order to save energy within the network, and periodic transmission of data is avoided. It has been observed and simulated by several researchers that for achieving best efficiency only 5% of nodes need to act as the CH.

The LEACH protocol operation is divided into two halves as the setup phase and the steady state phase. Organization of the cluster and selection of cluster head is done in the setup phase. Steady state deals with transfer of data from several nodes to the base station. To minimize the overhead during transmission, the duration of steady state is longer than the setup phase. During steady state, sensor nodes sense the environmental parameters and send data to the cluster head, where the cluster head aggregates the receiving data and sends it to the base station. After a set, defined time the network goes back to setup phase, where a new round of selection for the cluster head takes place. In order to reduce interference from node belonging to another cluster, the CDMA protocol is used for communication among clusters.

Power Efficient Gathering in Sensor Information Systems

PEGASIS is an enhanced version of the LEACH protocol which works on the concept of an optimal chain-based protocol, where the communication is set up with only the closest neighbors to extend the lifetime of network. Nodes takes turns to send information to the base station, thus round by round all nodes take turns to make communicate with the base station, and once one round end, another new round starts, and the process continues. This protocol also saves energy as the power required to transmit data per round is spread uniformly over all the nodes. This protocol, like LEACH, avoids formation of clusters and permits only a single node within a chain to transmit data to the base station.

In order to find the closet neighbor of a node within a network, PEGASIS uses signal strength to measure the distance such that only one node can be heard. It uses a greedy method for constructing the chain. The chain consists of near neighbor nodes forming the shortest path to the base station. The nodes in a chain aggregate the data to be sent to the base station. Thus the PEGASIS protocol focuses on two main objectives as increasing the lifetime of a node and optimizing the bandwidth consumption during communication.

Hierarchical Energy Aware Protocol for Routing and Aggregation in Sensor Networks

HEAP was proposed by Moazeni and Vahdatpour (2007) and is a hierarchical aggregation and routing protocol. This protocol is an energy-aware and fast route recovery protocol. Hierarchical aggregation is used for networks with close and intensive sensor nodes because network events can be sensed from several neighbor nodes, and iterative data packets can be produced at approximately the same time. Thus, aggregation can be used to remove the duplicity or redundancy of data and avoid high amounts of data transfer within the network and save energy. HEAP is governed by factors like message buffer, hierarchical, and routing level. The efficiency of this protocol can be increased by selecting the best values for these variables.

Hierarchical Periodic, Event-Driven and Query-Based

HPEQ works on a cluster-based hierarchical mechanism for transfer of data within the WSN. This protocol was proposed by Boukerche et al. (2004) and works by grouping sensor nodes into clusters for efficient relay of sensed data to the sink. In this protocol, there is uniform dissipation of energy among nodes along with a keen focus on reducing latency and network data traffic. A node with more residual energy is selected as the aggregator node for uniform transference of data to the sink.

Threshold Sensitive Energy Efficient Sensor Network Protocol

TEEN was proposed by Manjeshwar and Grawal (2001) wand was designed for time critical applications. This protocol provides a controlled tradeoff between data accuracy and energy efficiency by means of hard and soft threshold values.

Nodes within the network continuously sense environmental variables but transmission is done in a controlled manner not in a frequent manner in order to save energy, and this is achieved by means of attributes termed hard threshold and soft threshold values. Hard threshold is the threshold value of sensed attributes, whereas soft threshold is the small change in value of a sensed attribute that may trigger a node to switch on its transmitters and start transmitting. Thus, hard threshold reduces the number of transmission by allowing transmission only when sensed attributes are in the range of interest, whereas soft threshold further reduces the number of transmissions by restricting transmissions in cases of little or no change in sensed attributes. This is based on the idea that energy dissipation is more in transmission than in sensing the data.

Small Minimum Energy Communication Network

This protocol was proposed by Rodoplu and Meng (1999) and is a strategy to consume minimum energy within the network by making use of low power GPS devices. This protocol identifies a relay region that consists of nodes within a surrounding area where transmission through anode is more efficient as compared to direct transmission. SMECN finds a sub-network with fewer numbers of nodes requiring less power for transmission. This sub-network helps in sending messages on minimum energy paths. Thus, this protocol focuses on minimum energy consumption by a network.

Self-Organizing Protocol

This protocol proposed by Subramanian and Katz (2000) is used to support heterogeneous sensors within the network. This protocol can be used for mobile or static sensors. Sensors sense the environmental variables and forward the data to a centralized node which acts like a router among a designated set of nodes. The router node forms the backbone for communication and is static in nature. The routers should be reachable from each sensing node, and these routers transmit the data to the base station for communication. This protocol works in a hierarchical manner to form the communication path from sensing nodes to the base station.

Location-Based Routing

Location-based routing means nodes of the network are addressed by means of their location in the route. Incoming signal strength is used to find the distances between neighboring nodes. In order to find out exact locations of the nodes, several technologies like satellites and GPS can be used. Location information can be used to calculate the assumption of energy consumption within the network. In order to save energy within the network, the sensor node goes into sleep state in the case of no activity recorded. Some examples of location-based routing are minimum energy communication network (MECN), geographic adaptability fidelity (GAF), geographic and energy aware routing (GEAR), geographic distance routing (GEDIR), most forward within radius (MFR), and greedy other adaptive face routing (GOAFR). Discussion of some of these follows:

Geographic Adaptability Fidelity

This protocol was proposed by Xu et al. (2001) to increase the energy efficiency of a network. This protocol was primarily designed for mobile ad hoc networks but later was adapted for sensor networks. In this protocol, the network area is divided into fixed zones and virtual grids. Within a zone, each sensor contributes different tasks at different times such that any node is selected as an awake node which will go to sleep after a specified time. The awakened node will be responsible for monitoring and reporting of data to the BS on behalf of all the other nodes within the zone. Thus, by making other nodes sleep, this protocol saves energy. This task is performed by means of three different predefined states as *discovery*, which is used to determine the neighbor within the grid, *active*, which represent the nodes that are active for routing, and *sleep*, which determine the nodes that turn their radios off to save energy.

Geographic and Energy Aware Routing

This protocol proposed by Yu et. al. (2001) is an energy-aware protocol that uses geography to find geographic attributes and uses queries to find the appropriate selection of neighbors for packet routing. It works by region to form clusters in different regions, sending information to a particular region rather than the entire network and thus saves energy.

It works the on cost that is associated with each node, determined by a combination of residual energy and distance from the destination. This protocol works in two stages. First *forwarding the packet to target region* where a target node is selected based on the learning cost function and the distance from the node. Second *forwarding the packet within the region* where packets are diffused in the region by means of recursive geographic forwarding or restricted flooding.

Geographic Distance Routing

This is based on a greedy algorithm where packets are forwarded to a neighbor in the current vertex whose distance from the destination is minimal. This protocol was proposed by Stojmenovic and Lin (1999) along with protocols MFR and DIR based on basic localized routing algorithms.

Most Forward within Radius

Proposed by Takagi and Kleinrock (1984), this is the first position-based routing algorithm. In this protocol any packet forwarding to the destination will be forwarded to the next neighbor in the forwarding direction. This protocol introduces to define the most forward within the radius greedy routing algorithm.

Greedy Other Adaptive Face Routing

This protocol was proposed by Kuhn et al. (2003) and is a geometric ad hoc routing algorithm that merges a greedy and a face routing algorithm. This protocol selects the closet neighbor to which to forward the packet. It uses greedy forwarding and other adaptive face routing. The greedy forwarding approach is used for forwarding of packets to the destination, whereas greedy parameter stateless routing uses face routing in case of failure of greedy algorithm to find a route around dead-ends. The packet then reaches a node that is close to the destination node, where another adaptive face routing algorithm is used for recovery purpose (Table 5.1).

Protocol Operation

Different protocols under protocol operations follow.

Query-Based Routing

In the query-based routing protocol, the destination node generates a query in a natural language or high-level language for the data seeking task, and this query is sent to the entire network. Any node that has data that matches the query, sends the data in response back to the node. This query is matched with the table containing the sensed data from the sensor nodes. This routing protocol save energy by reducing energy consumption by means of performing data aggregation.

Multi-Path-Based Routing Protocol

This protocol deals uses multiple paths for transference of data from the source node to the destination node. This protocol increases the performance of a system by providing fault tolerance by means of providing an alternative path from the source to the destination in case of primary path failure. Regular periodic messages are used to keep these alternative paths alive. Network reliability increases at the cost of maintaining these live alternative paths.

Different authors have proposed algorithms under this protocol for minimizing power consumption and residual energy of the network. Li et al. (2001) proposed that residual energy of the route can be lowered by means of selecting a more energy efficient path. Similarly, other researchers provide solutions for energy efficiency for

Table 5.1 Criteria for Greedy Other Adaptive Face Routing

Criteria	Flat Routing	Hierarchical Routing	Location-Based Routing
Scheduling	Reservation-based	Contention-based	Sleep/awake-based
Collision	Avoided	Overhead present	Avoided
Duty cycle	Reduced duty cycle due to periodic sleeping	Variable duty cycle due to controlled sleep time	Periodic duty cycle with sleep and awake
Synchronization	Require global and local synchronization	Link formed without synchronization	Global synchronization
Latency	Lower latency due to multiple hop network formed by cluster head	Latency caused due to waking up of intermediate nodes and setting up of multiple paths	Low latency due to store and forward mechanism
Energy dissipation	Uniform and can be controlled	Depends on traffic pattern and adaptation to traffic patterns	Less as nodes goes to sleep when inactive
Fairness	Fair channel allocation	Fairness not guaranteed	Fairness not guaranteed
Efficiency	Overhead of cluster formation throughout the network	Routes form for region having data to send	Overhead of keeping location of each node

multi-path routing protocols. Thus, multi-path routing protocols provide an alternative for energy efficient recovery from failures in the WSN, and also the cost of maintaining these multi-paths is low.

Negotiation-Based Routing

Negotiation-based routing is based on the concept of avoiding propagation of duplicate packets. Flooding or gossiping may produce large amounts of data and may lead to receiving multiple data copies at a single node, leading to waste of resources and bandwidth.

This duplication of packets can be avoided by sharing a sequence of negotiation messages to all nodes to transmit redundant data to the next node. This reduces energy consumption and congestion within the network by reducing redundant data. SPIN protocol and its different versions are examples of negotiation-based routing protocol.

QoS-Based Routing

QoS-based routing is quality of service-based routing which maintains a tradeoff between quality of data and conservation of energy. In order to maintain QoS, certain parameters need to be taken care of such as energy, bandwidth, and delay. There are many protocols proposed in WSN that supports QoS and discussion of some of these follows.

Sequential assignment routing (SAR) was the first routing protocol that introduced the concept of the QoS routing protocol. The routing decision in SAR depends on three parameters: priority level of each packet, energy resources, and QoS on each path. This protocol also uses the multi-path approach for providing fault tolerance. For creating a multi-path, a tree is built from source to destination. The tree is built in such a manner that it avoids nodes with low energy for guaranteed QoS, but at the end of each sensor node is part of this tree. Thus, SAR works on the objective of minimizing the average weighted QoS metric throughout the lifetime of the WSN along with fault tolerance and easy recovery.

Coherent- and Non-Coherent-Based Routing

In order to process data, two different techniques can be used as coherent-based routing and non-coherent-based routing. Coherent-based routing works on the concept of sending minimum processed data to the aggregator, wherein aggregators are nodes that perform further processing. Minimum processing includes parameters like data suppression and time stamping. In non-coherent routing, nodes sends locally processed raw data to other nodes for further processing. Comparatively coherent routing provides for a better energy efficient routing protocol.

Data processing in non-coherent routing take place in three stages:

1. Detection of target, collection of data, and preprocessing of data
2. Declaration of membership
3. Election of central node

The first stage includes target detection, data collection, and preprocessing of data, whereas the second stage comes into the picture when nodes decide to participate in a cooperative function and declare their intention to neighboring nodes, so they have knowledge of the network topology as soon as possible. In the third stage, the central node is selected, and its work is to perform more sophisticated information processing. In order to undertake a more sophisticated task, this node needs to have a sufficient energy reserve and capacity for computational power.

Table 5.2 Summary of Routing Protocols

	Classification	Power Usage	Data Aggregation	QoS	Scalability	Multipath	Query-Based
SPIN	Flat	Limited	Yes	No	Limited	Yes	Yes
Directed diffusion	Flat	Limited	Yes	No	Limited	Yes	Yes
Rumor routing	Flat	N/A	Yes	No	Good	No	Yes
GBR	Flat	N/A	Yes	No	Limited	No	Yes
CADR	Flat	Limited	Yes	No	Limited	No	Yes
COUGAR	Flat	Limited	Yes	No	Limited	No	Yes
ACQUIRE	Flat	N/A	Yes	No	Limited	No	No
TEEN and APTEEN	Hierarchical	Maximum	Yes	No	Good	No	No
LEACH	Hierarchical	Maximum	Yes	No	Good	No	No
PEGASIS	Hierarchical	Maximum	No	No	Good	No	No
SMECN	Hierarchical	Maximum	No	No	Low	No	No
SOP	Hierarchical	N/A	No	No	Low	No	No
MECN	Location	Maximum	No	No	Low	No	No
GAF	Location	Limited	No	No	Good	No	No
GEAR	Location	Limited	No	No	Limited	No	No
GEDIR	Location	N/A	No	No	Limited	No	No
MFR	Location	N/A	No	No	Limited	No	No
GOAFR	Location	N/A	No	No	Good	No	No

Many algorithms have been proposed for coherent and non-coherent routing, such as the single winner algorithm (SWE) and multiple winner algorithm (MWE). Summaries of some of the protocols discussed above are in Table 5.2.

Research Issues with WSN Routing Protocol

With the progress and updating of routing protocols in WSNs, there are many research issues for future direction of WSNs. Discussion of which follow.

Secure mobile WSN routing: With the advancement in WSN technology, the need for implementation of WSN for mobile type application increases and securing routing protocol in such an application is also a future research directions.

Guaranteed security and QoS: For mission-critical applications, security and QoS are some of most important parameters to consider. Increasing security can degrade the performance of a WSN, so a proper tradeoff between security and QoS should be considered while designing WSN applications.

Protection of base station: The base station holds all the data from different sensors within the network, and in the case of mission-critical applications like fire, sensor node cooperation with the BS can be hampered. This issue should be kept in mind in future research directions along with security of the BS.

Understanding WSN protocol stack: For designing an efficient routing protocol, understanding the protocol stack is one of the key factors. The protocol stack consists of the physical layer, data link layer, network layer, transport layer, and application layer. The protocol stack contains plans for the task management plane, mobile management plane, and power management plane, so understanding of WSN protocol stack is crucial.

Multi-objective routing protocol: There are chances that while designing any routing protocol the consideration of some factors may lead to compromise of certain other attributes. For instance, designing a lightweight security mechanism can lead to compromise of security, as a secure protocol needs more processing power that could then increase congestion within the network. Designing a protocol that can change with the changing condition of the network is a research need.

Conclusion

This chapter focused on various parameters upon which protocols depend for efficient working within the WSN. This chapter also discussed various WSN protocols for various types of applications, with categorization and comparison to each other of their various attributes and factors. Open research issues are discussed to open a path for more research in these areas.

References

S. S. Al-Wakeel and S. A. Al-Swailemm, PRSA: A path redundancy based security algorithm for wireless sensor networks. In: *Proceedings of the IEEE Wireless Communications and Networking Conference* (WCNC 2007), 2007, pp. 4156–4160.

A. Boukerche, R. W. N. Pazzi and R. B. Araujo, A fast and reliable protocol for wireless sensor networks in critical conditions monitoring applications. In: *MSWiM '04: Proceedings of the 7th ACM International Symposium on Modeling, Analysis and Simulation of Wireless and Mobile Systems*, 2004, pp. 157–164.

J. Chu, P. D. S Dong, and G. Panganiban. Limb type-specific regulation of bric a brac contributes to morphological diversity. *Development* 129(3): 695–704, 2002.

S. K. Dhurandher, M. S. Obaidat, G. Jain, I. M. Ganesh and V. Shashidhar, An efficient and secure routing protocol for wireless sensor networks using multicasting. In: *Proceedings of the IEEE/ACM International Conference on Green Computing and Communications (GreenCom 2010) & IEEE/ACM International Conference on Cyber, Physical and Social Computing (CPSCom 2010)*, 2010, pp. 374–379.

R. Di Pietro, L. V. Mancini, L. Yee Wei, S. Etalle and P. Havinga, LKHW: A directed diffusionbased secure multicast scheme for wireless sensor networks. In: *Proceedings of the 32nd International Conference Parallel Processing Workshops (ICPPW '03)*, 2003, pp. 397–406.

N. El-Bendary, A. E. Hassanien, J. Sedano, O. S. Soliman and N. I. Ghali, mTESLA-based secure routing protocol for wireless sensor networks. In: *Proceedings of the first International Workshop on Security and Privacy Preserving in e-Societies*, 2011, pp. 34–39.

R. Fan, J. Chen, J. Q. Fu and L. D. A. Ping, Steiner-based secure multicast routing protocol for wireless sensor network. In: *Proceedings of the 2nd International Conference on Future Networks (ICFN 2010)*, 2010, pp. 159–163.

S. M. Gaurav, R. J. D'Souza and G. Varaprasad, Digital signature-based secure node disjoint multipath routing protocol for wireless sensor networks. *IEEE Sensors Journal* vol. 12, pp. 2941–2949, 2012.

W. B. Heinzelman, A. P. Chandrakasan and H. Balakrishnan, An application-specific protocol architecture for wireless microsensor networks. *IEEE Transactions on Wireless Communications*, vol. 1, pp. 660–670, 2002.

W. R. Heinzalman, A. Chandrakasan and H. Balakrishnan, Energy-efficient communication protocol for wireless microsensor networks. In: *Proceedings of the 33rd Hawaii International Conference on System Sciences*, 2000.

C. Intanagonwiwat, R. Govindan and D. Estrin, Directed diffusion: A scalable and robust communication paradigm for sensor networks. *Proceedings ACM MobiCom' 00*, 2000, pp. 56–67.

F. Kuhn, R. Wattenhofer and A. Zollinger, Worst-case optimal and average-case efficient geometric ad-hoc routing. In: *Proceeding of the 4th ACM Int. Symposium on Mobile Ad-Hoc Networking and Computing (MobiHoc)*, 2003.

Q. Li, J. Aslam and D. Rus, Hierarchical power-aware routing in sensor networks. In: *Proceedings of the DIMA CS Workshop on Pervasive Networking*, 2001.

W. Lou and Y. Kwon, H-SPREAD: A hybrid multipath scheme for secure and reliable data collection in wireless sensor networks. *IEEE Transactions on Vehicular Technology* vol. 55, pp. 1320–1330, 2006.

A. Manjeshwar and D. P. Grawal, TEEN: A protocol for enhanced efficiency in wireless sensor networks[C]. In: *Proceeding of the 15th Parallel and Distributed Processing Symposium*, 2001, pp. 2009–2015.

M. Moazeni and A. Vahdatpour, HEAP: A hierarchical energy aware protocol for routing and aggregation in sensor networks. In: *Proceedings of the 3rd International Conference on Wireless Internet*, 2007.

V. Rodoplu and T. H. Meng, Minimum energy mobile wireless networks. *IEEE Journal on Selected Areas in Communications*, vol. 17, no. 8, pp. 1333–1344, 1999.

N. Sadagopan, B. Krishnamachari and A. Helmy, The ACQUIRE Mechanism for Efficient Querying in Sensor Networks. In: *Proceedings of the First IEEE International Workshop on Sensor Network Protocols and Applications (SNPA)*, 2003, pp. 149–155.

C. Schurgers and M. B. Srivastava, Energy efficient routing in wireless sensor networks. In: *The MILCOM Proceedings on Communications for Network – Centric Operations: Creating the Information Force*, 2001.

I. Stojmenovic and X. Lin, GEDIR: Loop-free location-based routing in wireless networks. In: *Int'l. Conf. Parallel and Distrib. Comp, and Sys.*, 1999.

W. Su and I. Akyildiz, A stream enabled routing (SER) protocol for sensor networks, *Medhoc-Net*, 2002.

L. Subramanian and R. H. Katz, An architecture for building self-configurable systems. In: *IEEE/ACM Workshop on Mobile Ad Hoc Networking and Computing (MobiHOC 2000)*, 2000.

H. Takagi and L. Kleinrock, Optimal transmission ranges for randomly distributed packet radio terminals. *IEEE Transactions on Communications*, vol. 32, no. 3, pp. 246–257, 1984.

A. D. Wood, L. Fang, J. A. Stankovic and T. He, SIGF: A family of configurable, secure routing protocols for wireless sensor networks. In: *Proceedings of the 4th ACM Workshop on Security of Ad Hoc and Sensor Networks*, 2006, pp. 35–4.

Z. Xizheng, Survivable and efficient clustered keying in wireless sensor network. In: *Proceedings of the 8th ACIS International Conference on Software Engineering, Artificial Intelligence, Networking, and Parallel/Distributed Computing (SNPD 2007)*, 2007, pp. 688–693.

Y. Xu, J. Heidemann and D. Estrin, Geography-informed energy conservation for ad hoc routing. In: *Proceedings of the 7th Annual International Conference on Mobile Computing and Networking (MobiCom '01)*, 2001, pp. 70–84.

Y. Yao and J. Gehrke, The cougar approach to in – network query processing in sensor networks. In: *SIGMOD Record*, 2002.

Y. Yu, R. Govindan and D. Estrin, Geographical and energy aware routing: A recursive data dissemination protocol for wireless sensor networks. UCLA Computer Science Department Technical Report. 463, 2001.

Chapter 6

Routing Schemes in Wireless Multimedia Sensor Networks

Introduction

There are continuous advancements in research and usage of wireless sensor technology with respect to the types of applications being implemented. Wireless sensor networks can make use of multimedia devices like microphones and video cameras for applications. These small and low cost devices can be integrated with a wireless sensor network and used for multiple applications. Integration of wireless sensor networks with these multimedia devices leads to a special type of wireless sensor network called a wireless multimedia sensor network (WMSN). This network is capable of sensing and sending multimedia and scalar data like audio, video, and images in real-time, as well as non-real-time systems. A WMSN network can be applicable in many situations like avoidance of traffic by means of a camera-based system, multimedia surveillance system, monitoring of environment by means of sensors and multimedia devices, and control and enforcement system.

While designing a WMSN, there are various factors that need to be considered and kept in view as follows:

Bandwidth requirement: Multimedia content like videos, etc. requires high bandwidth as compared to normal environmental variables. To facilitate WMSN a ultra-wideband transmission (UWB) technique is promising for the transfer of data in a WMSN network.

Quality of service: The quality of service (QoS) requirement is mandatory for a WMSN network, as there are many multimedia data that need to be transferred and need high quality resolution and bandwidth. For example, multimedia data like snapshots contain event-trigger observation that needs to be obtained in a short span of time. Similarly, streaming multimedia data is generated over a longer time period and requires sustained information delivery. So these types of deliveries require high QoS within the WMSN network.

Multimedia source code techniques: In order to maintain high bandwidth for a WMSN network, an advanced source code technique is required. Traditional source code techniques are based on reducing bit rate generation from the source which can be achieved by means of reducing intra-frame compression. The latest techniques and algorithms used are based on distributed source coding, where source statistics are exploits at decoder. Design of such types of techniques and algorithms is needed for efficient bandwidth and QoS delivery for a WMSN network.

Power consumption: Since the sensors used in a WMSN network generate a high volume of data, and these sensors are battery operated, there will be a need for high power for communication in WMSN networks. While designing algorithms and protocols for WMSN networks, reducing power consumption is one of the primary concerns.

Integration with internet: Most of the multimedia and WSN applications need to be accessed from anywhere and everywhere. This type of integration requirement for multimedia sensors with internet architecture for availability of data anywhere and everywhere is a future need.

Architecture of WMSN: WMSN architecture can be divided into several classes based on the type of sensors/nodes being used. These classifications follow:

Single-tier flat and homogenous: In this type of architecture, sensors with the same physical capabilities are used. It uses distributed processing with centralized storage for processing of data.

Single-tier clustered and heterogeneous: In this type of architecture, sensors with different physical capabilities are used. In this architecture nodes have different capabilities, different processing speed, and batteries. All the data are sent to a centralized node with more capacity in terms of power, processing, etc. termed the cluster head (CH) that processes and transfers information to the sink node for further processing.

Multitier heterogeneous: In this type of architecture, multiple layers of nodes are set up with different speeds and processing tasks at each layer. This architecture uses distributed processing at each layer. Data stored in a distributed fashion in this architecture is sent to a base station for processing (Akyildiz et al., 2007).

Layered architecture of WMSN consists of the following layers:

Application layer: This layer provide services for traffic management and admission control functionality by preventing applications establishing data flows when network resources are needed elsewhere and not available. This layer also performs coding of the source as per the requirements of the application and constraint of the hardware, using advanced techniques of multimedia encoding. This layer also provides an efficient and flexible system software for exporting services for higher-layer applications to build on. The application layer also provides primitives for applications to leverage collaborative, advanced in-network multimedia processing techniques.

Transport layer: This layer provides end-to-end reliability and support for congestion control within the network. The functionality and working of the transport layer can be categorized into two parts as TCP/UDP and TCP friendly schemes for WMSN and application-specific and non-standardized protocols. In a WMSN network, UDP is preferred in the case of real-time applications like streaming of videos, as timeliness in this case is of greater importance than reliability. In rest of multimedia cases, TCP is preferred over UDP for providing reliable service. Application-specific and non-standard protocols are divided into three subcategories based on the types of applications into reliability, congestion control, and multipath. As all these three conditions are important factors for designing the WMSN network at transport layer.

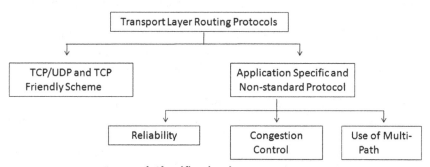

Protocol Classification in Transport Layer

Network layer: The network layer is responsible for providing QoS for a WMSN network. QoS is one of the necessary parameters for providing efficient service in a WMSN network for high bandwidth, video, audio, and data. Just like traditional networking, the network layer supports IP addressing and routing protocol. Where routing protocol in the case of a WMSN network is classified into three subcategories as *network-condition-based, traffic-classes-based, and real-time streaming–based*. Network-condition-based routing protocol is based on metrics like position with respect to sink, radio characteristics, rate of error, backlogged packets, and residual energy. Traffic-classes-based routing protocol depends upon metrics like QoS and traffic classes, rate of dropping of packets, latency tolerance, and bandwidth requirements. Real-time streaming–based routing protocol depends on metrics like spatio-temporal characteristics and probability of delay guarantee (Akyildiz et al., 2007).

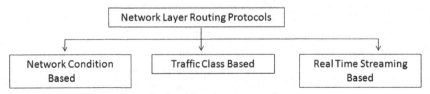

Protocol Classification in Routing Layer

MAC layer: The MAC layer is responsible for providing reliable, error-free data transfer with minimum retransmission for particular applications. The MAC layer also supports QoS for high applications which may need video, audio, etc. QoS within the MAC layer can be classified into following three subparts, channel access policies, scheduling and buffer management, and error control. Thus, protocols in the MAC layer can be broadly categorized into two sections as *contention-free* and *contention based*. Contention free can be further classified into *single channel* and *multiple channel* based. For single channel time division multiple access (TDMA) is the representative protocol for this class, where a sink or cluster head is responsible for slot assignment. Single channel protocol gives better control for multimedia design parameters and having simple hardware and operation as compared to other protocols. This protocol is based on multiple input and multiple output (MIMO) technology. Examples of single channel protocol are STE and EDD. Multichannel provides better bandwidth utilization. In the case of multichannel, channel switching delay must be considered in end-to-end latency. Examples for multichannel protocol are STEM, RATE-EST, and CMAC. Contention-based protocol for MAC layer works by means of sleep and awake cycles. It uses a burst for scheduling that may lead to jitter within the network. Examples of contention-based protocol are S-MAC and T-MAC.

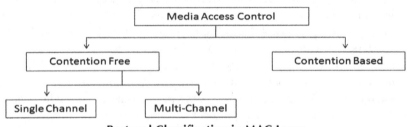

Protocol Classification in MAC Layer

Physical layer: The physical layer of a WMSN is responsible for selection of frequency, modulation, and channel encoding. This layer consists of basic hardware transmission technologies. Physical layer technology can be classified into three subgroups as *ultra-wide band, spread spectrum* and *narrow band* based on bandwidth consideration and modulation scheme. Classification can also be made to different standard protocols such as Zigbee, IEEE 802.15.1, IEEE 802.15.4, Bluetooth, IEEE 802.11, and 802.15.3a UWB. These protocols differ in terms of their data rate, output power, frequency, and number of nodes as shown in the following table.

Criteria	UWB	Zigbee	Bluetooth	IEEE 802.11
Data rate (Max.)	250 Mbps	250 Kbps	3Mbps	54 Mbps
Output Power	1 mW	1-2 mW	1-100 mW	40-200 mW
Range	< 10 meters	10-100 meters	1-100 meters	30-100 meters
Frequency	3.1- 10.6 GHz	2.4 GHz or 915 MHz or 868 MHz	2.4 GHz	2.4 GHz
Code efficiency	97.94 %	76.52 %	94.41 %	97.18 %
No. of nodes	-	<65000	7	30

Standards Specification in Physical Layer

WMSN Routing Protocols

There are various requirements for WMSN routing protocols that govern the selection of routing protocol for various types of applications. These parameters govern the performance of WMSN routing protocols in terms of energy efficiency and quality of service. The following section discusses these parameters.

QoS requirements: Quality of Service (QoS) is needed for WMSN due to the type of data using the WMSN. WMSNs carry data like video and audio that need high bandwidth and efficiency by means of high QoS. Various parameters that govern QoS within a networks are *bandwidth, latency, reliability, jitter* and *lifetime*. High bandwidth is required by multimedia traffic used in a WMSN, and this can be achieved by optimizing the resources and by means of employing multipath, multichannel, and stream division methods. Latency is the end-to-end delay that can be classified into soft latency bounded system and hard latency bounded system (Akyildiz et al., 2007). A soft latency bounded system has to maintain probabilistic delay guarantee, whereas a hard latency bounded system has to maintain deterministic latency. Reliability means delivery of data to the destination with minimum losses caused due to congestion or interference, and this is one of the important factors for achieving QoS. Jitter is accepted variability of delay between received packets which if not met may result in error, glitches, or discontinuity in the video or audio stream. Lifetime is the duration of sustainability of a network.

Energy efficiency: Since the data used in a WMSN requires high bandwidth it draws more power from sensor nodes that operate on battery. Designing a protocol which is energy efficient is one of the major parameters that needs to be considered.

Traffic classes: In order to provide differentiated services among different applications used in WMSN, QoS can be designed by means of defining various traffic classes based on different criteria. These traffic classes can be defined as follows:

Real-time, loss-tolerant multimedia streams: This class includes video stream, audio stream, multilevel stream, scalar data, and multimedia. Due to such traffic, this class requires high bandwidth and strict delay bounds.

Delay-tolerant, loss-tolerant multimedia streams: This class includes offline storing and processing traffic that must be transmitted in real-time to avoid any packet loss.

Delay-tolerant, loss-tolerant scalar data: This class includes data from any scalar sensor, non-time critical multimedia snapshot. Bandwidth requirement for such data is low.

Delay-tolerant, loss-intolerant scalar data: This class of traffic includes data from monitoring processes and data intended for offline storing or processing. This type of data requires less bandwidth.

Real-time, loss-tolerant scalar data: This class includes data from monitoring processes or loss tolerant snapshot multimedia. Bandwidth demand for this data is low.

Real-time, loss-intolerant scalar data: This class includes data from monitoring processes that are time critical. Bandwidth demand for this class is low (Ehsan and Hamdaoui, 2012; Akyildiz et al., 2007).

Hole detection: Holes are dynamic in nature and are created due to exhaustion of energy of the routes that are made due to high demand of bandwidth and low end-to-end delay for real-time applications. These holes can be avoided by means of designing a protocol which facilitate balanced energy usage within the network (Ehsan and Hamdaoui, 2012).

Classification of WMSN Routing Protocol

Routing protocols are used for transferring data from source to destination in a secure and efficient manner. Selection of a suitable routing protocol is one of the main concerns in WMSN protocols, as these protocols require high bandwidth and throughput due to type of data (video and audio). Based on different classifications and attributes values, routing protocols for WMSN are divided into multiple categories as listed below. These protocols are categorized into different categories based on attributes like network types, suitability with respect to WMSN, performance of the protocol, and energy efficiency. The classes and subclasses of these routing protocol classifications are not mutually exclusive, as they may belong to more than one class or subclass. Broadly speaking, WMSN routing protocols are divided into three class as *QoS based* or *traffic class based* which are further subdivided into *latency constraint* and *multi-constraint*, and these are further subdivided into *hard real-time* (HRT), *soft-real-time* (SRT) and *firm real-time* (FRT) based on factors like latency constraint and deadline delay. Another category of classification of WMSN routing protocol is *swarm intelligence* or *real-time streaming based*. A third category is based on the network structure of WMSN further classified into *flat routing*, *hierarchical routing* and *location based routing*.

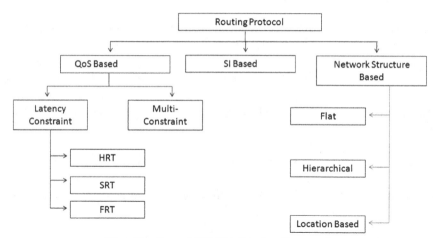

Classification of WMSN Routing Protocols

QoS-based routing protocol: QoS within a WMSN routing protocol can be achieved by means of parameters like jitter, bandwidth, latency, and reliability. A QoS based routing protocol can be further subdivided into latency based and multi-constraint based on the number and types of constraint in this protocol.

Latency constraint-based routing: This protocol is further subdivided into HRT, SRT, and FRT based on latency constraint and deadline delay. Where HRT is a hard real-time constraint that considers a system failure if packets arrive after a predefined delay deadline. In a real-time system, maintaining such a hard constraint with factors like error prone medium and loss of links is very difficult. SRT systems are soft, real-time-based systems which allow some packets to be lost from the deadline but guaranteed for a probabilistic delay in order to a avoid system crash. A FRT system uses the concept of (m, k) that says that at least m out of k consecutive messages must be delivered prior to their delay deadline.

HRT-Based Protocol

The protocol proposed by Zhou and Wang (2010) is a delay constraint routing protocol that focused on a hard, real-time-based guarantee and also minimizing energy consumption in the WMSN. This protocol uses a global link-based routing protocol which demands the knowledge of the entire network for finding the best route through the network. The working of this protocol is divided into three subgroups as the first phase remove nodes having less residual energy from group to avoid any hole problems. The second phase implements Dijkstra's algorithm for finding the best route. The third phase ensures that all paths meet the delay upper bound. On failing the upper bound limit, Dijktra's algorithm finds the alternative best path. Thus, this protocol focuses on QoS and energy efficiency of the network and guarantees a hard-real-time delay bound. This protocol is suitable for smaller networks. For larger networks, there is large energy, bandwidth, and memory utilization for the information exchange between nodes.

SRT-based protocol: The SRT-based protocol involves many protocols such as sequential assignment routing (SAR), energy-aware QoS (EAQoS), real-time communication architecture for large-scale WSN (RAP), stateless protocol for real-time communication (SPEED), real-time power-aware routing (RPAR), optimized energy-delay sub-network routing (OEDSR), service differentiated real-time communication scheme (SDRCS), adaptive greedy-compass energy-aware multipath (AGEM), energy efficient and QoS-aware multipath routing (EQSR). Discussion of these protocols follows.

Sequential assignment routing (SAR): This protocol was proposed by Sohrabi et al. (2000). SAR was the first routing protocol that introduced the concept of QoS to a routing protocol. Routing decisions in SAR is depend on three parameters to determine the priority level of each packet, energy resources, and the QoS on each path. SAR uses single winner election (SWE) and spanning tree (ST) algorithms for finding the minimum hop spanning tree. This protocol also uses a multipath approach for providing fault tolerance. For creating multipath, a tree is built from the source to destination. The tree is built in such a manner that it avoids nodes with low energy for guaranteed QoS, but the end result is that each sensor node will be part of this tree. Thus, SAR works on the objective of minimizing the average weighted QoS metric throughout the lifetime of a WSN along with fault tolerance and easy recovery. This protocol lacks supporting scalability of the network.

Energy-aware QoS (EAQoS): This protocol was proposed by Akkaya and Younis (2003) and works on the concept that decreasing energy consumption will increase the lifetime of a network. This is an energy-aware QoS routing protocol proposed for data on a WMSN network for the efficient running of best-effort traffic. This protocol finds the least costly, delay constraint path for real-time traffic on a WMSN in terms of link cost capturing nodes' energy reserves, transmission energy, rate of error, and other communication parameters.

This protocol can be used for real-time and non-real-time traffic and is based on a class-based queuing model. This protocol uses an advanced version of Dijkstra's algorithm that works based on a least cost model, where cost is associated with each link that is stored. For every path discovered, a tree like structure is created and the total cost for the path is calculated, provided only those paths are chosen that meet the minimum criteria for QoS requirement and maximum network throughput. This is a cluster-based energy efficient and delay constraint routing protocol that guarantees bandwidth and energy efficient paths and is also used for reducing congestion over newly found paths.

Real-time communication architecture for large-scale WSN (RAP): RAP is a localized routing protocol based on priority and proposed by Lu et al. (2002). This protocol proposes a new scheduling policy called velocity monotonic scheduling (VMS) that schedules packets based on requested velocity. It is also a location-based protocol, as it assumes that each node within a WMSN network knows about its location, and this combination of VMS with geographic forwarding (GF) make this protocol scalable and suitable for multimedia data. This protocol also supports query driven and event-oriented services.

Stateless protocol for real-time communication (SPEED): This is a stateless, localized protocol that supports QoS with minimum MAC layer support, and proposed by He et al. (2003). This protocol uses a stateless nondeterministic geographic forwarding (SNGF) method for routing within the network and maintains the required delivery speed and traffic balance for reducing congestion within the network. SNGF uses four modules as a beacon exchange, delay estimation, neighborhood feedback loop, and beck pressure rerouting for setting up communication within the network. This protocol avoids void problems and supports load balancing.

Real-time power-aware routing (RPAR): This protocol proposed by Chipara et al. (2006) achieves application-specific delay requirements by dynamically adjusting transmission power and other routing metrics. RPAR works in four components as neighborhood manager, delay estimator, forwarding policy, and velocity assignment policy. Where neighborhood manager finds a route for forwarding by meeting the demand of end-to-end delay and energy efficiency for the network. This protocol should be used for real-time traffic only. It uses Jacobson's algorithm for finding the transmission count and estimation delay of the network and for calculating retransmission timeout. RPAR also supports power adaptation, neighbor discovery, and handling of congestion holes. This is a scalable routing protocol that supports scalable network.

Optimized energy-delay sub-network routing (OEDSR): This protocol was proposed by Ratnaraj et al. (2006) and is an event-driven and cluster-based routing protocol where the information dissipation from the cluster head (CH) to the base station (BS) is multi hop. It searches for the best route based on parameters like end-to-end delay, available amount of energy, and distance from the BS.

This protocol selects a node with maximum energy as the temporary head (TH) whose work is to select and calculate the number of CHs within a cluster. After this selection, TH broadcast this information to all nodes and then becomes a simple or regular node. It is a multipath and energy efficient protocol that selects loop-free routes for data transmission and provides suitable solutions for congestion within the network.

Service differentiated real-time communication scheme (SDRCS): This protocol was proposed by Xue et al. (2011) and is an event driven routing protocol that implements cross layer functionality for routing of real-time traffic. This protocol uses the integration of dynamic forwarding techniques and CSMA/CA based prioritization techniques. This protocol uses a polling contention period and signal strength-based grouping for scheduling of packets. This protocol is divided into five subsections as follows: RSS-based grouping, admission control, prioritized queuing, real-time MAC, and dynamic forwarding. Grouping granularity is used for maintaining distance accuracy. The First in first out (FIFO) technique is used for data scheduling. Per-hop deadline-based queuing is used for the queuing policy. For localization there is no need for any extra hardware. This protocol can adapt to network dynamics like channel void or channel quality.

Adaptive greedy-compass energy-aware multipath (AGEM): This protocol proposed by Medjiah et al. (2010) is a location-based protocol that has a load balancing feature for increasing the life time of the network and also to decrease the queue size at nodes. This protocol uses an adaptive compass policy for selecting subsets of neighbors. The forwarding policy of this routing protocol is governed by four parameters: hop-count, remaining energy, distance of current node to its neighbor, and history of forwarded packets in the same stream. This protocol works in two phases as smart greedy forwarding and walking back forwarding. The location of nodes is detected by GPS or by using distributed localized techniques. Each node maintains a database of its adjacent neighbors. This protocol is a multipath routing protocol that maintains load balancing and can be used for static and mobile-based sensor networks.

Energy efficient and QoS-aware multipath routing (EQSR): This protocol was proposed by Ben-Othman and Yahya (2010) and it increases the lifetime of a network by minimizing energy consumption rate. Routing is through parameters like available buffer size, residual energy, and signal-to-noise ratio. This protocol can be used for real and non-real type traffic by implementing a queuing model. This protocol makes maximum use of available network resources by discovering node-disjoint paths. A path is discovered through three phase as the initialization phase, primary path setup phase, and alternate path setup phase. To determine the preferred next hop link, the cost function can be used. This protocol updates its pre-established path as and when the values of metrics are updated.

The packets of this protocol are segmented into equal size segments that are scheduled for transmission. Before packet scheduling, an error correction code is added through the traffic allocation scheme that helps in recovering from node failure. A weighted traffic allocation scheme is used for distribution of data along available paths. This is an energy efficient multipath routing protocol that is capable of handling real- and non-real-time traffic.

FRT-Based Protocol

FRT-based protocol was proposed by Li and Kim (2014) and uses the concept of (m, k) which says that at least m out of k consecutive message must be delivered prior to their delay deadline. This protocol uses a real-time fault tolerance mechanism of local status indicator (LSI) for achieving application QoS requirements. LSI accesses the local condition of nodes and detects any sort of fault within the network. This protocol is divided into four phases: beacon exchange among neighbors, estimation of single hop delay, calculation of LSI, and fault recovery mechanism.

Periodic beaconing is used to share local information among neighbors and increase the lifetime of an overloaded node and the whole network. This beacon contains information for location and the residual energy of nodes. Three types of beacons are used for sharing information with different functionality: delay estimation beacon, stream DBP beacon, and orphan node removal beacon. This protocol is very robust and can fulfill required latency and reliability requirements. As this protocol uses local system evaluation, it reduces routing overhead and avoids link failure.

Multi Constraint

The multiple QoS constraint routing protocol focuses on parameters like latency constraint, reliability, load balancing, bandwidth, and congestion. There are many protocols proposed under this section as multipath multi-SPEED (MMSPEED), distributed aggregate routing algorithm (DARA), real-time routing protocol with load distribution (RLTD), latency and reliability-aware geographic routing (LRGR), QoS-aware multi-sink opportunistic routing (QMOR), and multi constrained routing algorithm (MCRA).

Multipath Multi-SPEED (MMSPEED)

This protocol was proposed by Felemban et al. (2006) and is a location-based protocol that uses a multipath forwarding mechanism to increase the reliability of the network. As the nodes of the network are aware of their locations, this protocol supports scalability and self-adaptiveness, and it also removes setup and recovery in the network. This protocol provides service differentiation by virtually isolating the speed layer. This virtual isolation is based on speed classes and placement of these speed classes into a priority queue. Probabilistic multipath forwarding is used to differentiate packets with different requirements. Thus, this protocol saves energy by making proper use of paths. MMSPEED uses back pressure to remove void area problems and reduces incoming traffic.

The MMSPEED protocol is a location-based and query-driven protocol that operates with the goal of providing timely QoS in a reliable domain with localized packet routing decision making. By making use of redundant paths, it prevents unnecessary flooding to achieve load balancing, network lifetime, and improved reliability.

Distributed Aggregate Routing Algorithm (DARA)

DARA was proposed by Razzaque et al. (2008) and is a multipath, multi-sink and location-based protocol with the objective of improving the reliability of paths and guaranteeing the delay by preserving the battery power of the nodes. DARA is a fully localized protocol which removes the need for maintaining any global state information. DARA supports for dynamic and static types of networking topologies. It increases the flexibility of the network by sending duplicate packets from the source node. For giving priority to real-time traffic, it modifies its MAC table. It uses a power control transmission scheme and reduces retransmission to reduce energy consumption on the network.

Real-Time Routing Protocol with Load Distribution (RLTD)

The RLTD protocol was proposed by Ahmed and Fisal (2008) and ensures high throughput and prolongs network lifetime. RLTD is a location-based routing protocol which combines the functionality of geocast forwarding with link quality. It achieve delay bounds by maximizing velocity and the remaining energy of the nodes. It can detect and bypass holes and also supports for query-based services and load balancing. RLTD is divided into four phases: location management, routing management, power management, and neighborhood management. Location management uses the location of predetermined neighboring nodes to calculate the location of all nodes. Routing management is used for making routing decisions and computing the optimal forwarding path. Power management is used to determine the state of transceivers and transmission power of sensor nodes, and neighborhood management is used for discovering the subset of forwarding candidate nodes and maintaining forwarding to neighbor table.

Latency and Reliability-Aware Geographic Routing (LRGR)

LRGR protocol was proposed by Rao et al. (2013) and is a hierarchical routing protocol focused on prediction of mobility and energy-aware cluster formation. LRGR is divided into multiple rounds where each round is associated with two phases, the clustering phase and the routing and data transmission phase. The clustering phase is further subdivided into energy exchange, CH set up, and cluster formation. In the clustering phase, all nodes share their residual energy with all other nodes, and average residual energy is calculated by each node. The node with the highest ratio of residual energy to average residual energy is selected as the cluster head. The most stable path is selected based on parameters like connectivity, balanced energy usage, position, and congestion.

QoS-Aware Multi-Sink Opportunistic Routing (QMOR)

The QMOR protocol was proposed by Shen et al. (2014) and is designed for real-time multimedia traffic with an objective of minimizing energy consumption and maintaining the delay and reliability requirements of the network. QMOR is a multi-sink routing protocol which uses an opportunistic routing scheme for reducing energy consumption. It make use of priority-based forwarding for implementing QoS. For maintaining the QoS requirement, the QMOR protocol selects and prioritizes the forwarder list. The QMOR protocol reduces redundancy and maximizes energy gains by using a correlation-aware differential coding scheme. It uses a differential coding-based source and intermediate selection (DCISIS) protocol for finding the source node and destination node within the same correlation group.

Multiconstrained Routing Algorithm (MCRA)

The MCRA protocol was proposed by Intanagonwiwat et al. (2003) and is a location-based multi constrained protocol supporting query-based services. For querying any event, a special message called "interest" is used. It uses a logical coordinate system for calculating the coordinates of the node by means of hop-count information which reduces the message overhead. It uses message suppression techniques for reducing redundant transmission and retransmission of messages caused due to collisions. Messages are suppressed through restraining or deferring forwarding.

SI-Based Routing Protocol

This protocol is influenced by the collective behavior of biological species like ant colonies which are intelligent in nature. These biological species work in coordination with each other, and their problem-solving method was used for designing this protocol. It makes use of mobile software agents for simulating the behavior of these species. The SI-based routing protocol is autonomous, fast, adaptive, scalable, modular, and fault tolerant in nature. There are various protocols designed on the behavior of biological species like distributed stigmergic control for communication networks (AntNet), multimedia-enabled improved adaptive routing (M-IAR), Ant-based service-aware routing (ASAR), Ant colony optimization-based load-balancing routing algorithm (ACOLBR), ant colony optimization-based QoS routing algorithm (ACOWHSN), AntSenNet, hop- and load-based energy-aware routing protocol (HLEAR), AntHQSeN, an agent-assisted QoS-based routing (QoS-PSO), a QoS-aware routing algorithm based on ant-cluster (ICACR), feedback-enhanced learning approach (FROMS), strength pareto evolutionary algorithm (SPEA), bee sensor, bee sensor –C, (EQR-RL), and FTIEE.

AntNet: This protocol was proposed by Di Caro and Dorigo (1998) and is an adaptive and hybrid routing protocol which uses the ACO algorithm and reinforcement learning for finding a route within the network for transmission of multimedia data. It is a multipath, multi constrained and multi objective protocol routing protocol which is stochastic in nature and delivers the data in real-time-latency constraint. It make use of agents called forward ants and backward ants for implementing adaptive learning of the routing tables. Forward ants are send at regular intervals from any network node to the destination for discovering a low-cost path and feasibility of the network. For calculating forward probability, QoS parameters such as delay, bandwidth, and residual energy are used to make processes adaptable to traffic distribution.

Multimedia-enabled improved adaptive routing (M-IAR): M-IAR was proposed by Rahman et al. (2008) is a localized and ant-based routing protocol which focuses on end-to-end delay, jitter, and energy efficiency. This is an ant-based flat multi hop routing protocol which is self-organized, fault-tolerant, and adaptive in nature. This protocol considers the closeness of next-hop neighbor from the sink, along with the distance of the neighbor from the sender node. It uses a piggybacking technique for acknowledgement to save bandwidth of the link.

Ant-based service-aware routing (ASAR): ASAR was proposed by Sun et al. (2008) and is a clustered, ant colony optimization-based routing protocol. ASAR is an ant-based service-aware hierarchical protocol that supports query driven as well as data driven services. For accelerating the convergence of algorithms and optimizing network resources, it make use of the concept "pheromone quantization". This algorithm runs in all CHs and makes use of positive feedback to find three different routes for each type of services, and the results are stored in CHs. For routing data it makes use of network, bandwidth, packet loss rate, delay, energy consumption, and network lifetime parameters. ASAR supports three types of services and meets diverse QoS requirements.

Ant colony optimization-based load-balancing routing algorithm (ACOLBR): The ACOLBR protocol was proposed by Bi et al. (2010) and is a hierarchical, swarm-based, and multipath-based routing protocol which focuses on congestion control. ASAR creates minimum a spanning tree by using a spanning tree algorithm. This is a hierarchical routing protocol that is further sub-divided into intercluster routing and intracluster routing. Where intracluster makes use of a minimum spanning tree with a CH as the root node, intercluster uses the ACO algorithm to find the minimum cost and optimal path from the source to the destination. ACOLBR enhances the exploring ability of ants by introducing maximum and minimum values of pheromone density. ACOLBR can recover from path failure by setting pheromone values to zero and then sending an error message to the sender. If the amount of data is greater than the maximum path flow, then this protocol splits streams to improve t transmission.

Ant colony optimization-based QoS routing algorithm (ACOWHSN): This protocol was proposed by Yu et al. (2011) and is an ant colony-based routing protocol which is an energy-aware and QoS-based protocol. This protocol is reactive in nature and finds a route only when required in order to save energy. This protocol is also adapts to any change within the network. It selects routes which meets application-specific QoS requirements. ACOWHSN uses forward ants and reverse ants for setting up communication within the network. Forward ants store information on nodes visited, minimum residual energy, cumulative queue delay, and packet loss to be used for evaluation of routes based on QoS. It make use of heuristics for finding the minimum cost route. Routing is performed in two stages as route discovery and routing confirmation. In routing discovery forward ants are sent towards the destination node using a broadcast or unicast approach, and if there is no information for the destination node then broadcasting can be used. In the routing confirm stage, forward ants are converted into backward ants that travel in a backward direction from the destination to the source to update the pheromone table of every intermediate node. The ACOWHSN protocol increases network lifetime by satisfying delay, bandwidth, and packet loss.

AntSenNet: This protocol was proposed by Cobo et al. (2010) and is a hybrid routing protocol comprised of reactive and proactive components. It makes use of the ACO-based routing protocol and clustering approach for transference of multimedia data. For minimizing distortion in the transmission of video, it supports power efficient multipath video packet scheduling. AntSenNet is a complete distributed clustering algorithm which supports different types of traffic classes. It is a scalable and multipath routing protocol which can also handle congestion problems.

The operation of this protocol is divided into rounds and is further divided into t cluster setup phase and steady phase. Cluster ants (CANTS) are agents that are used for setting up the clusters. For regularly updating table attributes like ID, clustering pheromone value, and its state, HELLO packets are broadcast in regular intervals. The clustering pheromone value is used to determine whether the current node is suitable as a CH. The cluster radius is used for finding the minimum distance between two CH nodes.

Updating the routing table is done in three phases as forward ant phase, backward ant phase, and route maintenance phase. Forward ants are generated when there is no satisfactory route to the sink node, so the forward node carries the information of route they travel and elevates the route as they reach the sink node. In the case of an elevated route unable to meet application-specific requirements, then a backward route from sink to node is established by means of a backward ant. The maintenance phase maintains the congestion and link loss problems of the route and network.

Hop- and load-based energy-aware routing protocol (HLEAR): HLEAR was proposed by Nayyar et al. (2011a,b) and is an energy efficient, swarm-based reactive protocol that avoids holes and increases network lifetime. HLEAR uses parameters like hop-count, residual energy, and rate of forwarding nodes for finding the best path. It is divided into two phases: route discovery and path establishment. In route discovery, RREQ packets are broadcast to the entire network in a controlled manner. In path establishment, a reply message RREP is sent back to set up the communication. HLEAR finds the minimum distance path by finding the disjoint path. This protocol uses hop-count as its metric.

AntHQSeN: The AntHQNet protocol was proposed by Kumar et al. (2014) and is designed for a heterogeneous sensor network. It is a reactive ant-based routing protocol which can be used in static and dynamic topologies. This protocol treats scalar and multimedia traffic differently, and it finds a consistent solution by means of a reactive route discovery mechanism.

This protocol can be divided into two phases: route discovery and route maintenance. In route discovery phase, agents called forwarding ants are created by a source node for finding multiple paths to the destination node. These nodes carry information on the network for evaluating various parameters of QoS. The next node is determined by the forwarding node based on pheromone values that make the protocol adaptable to changes in the network. The route maintenance phase is used in the case of node failure. Hello ants' messages play an important role in this protocol for neighbor discovery, exchanging information about the bandwidth, timestamp, energy, and pheromone concentration.

An agent-assisted QoS-based routing (QoS-PSO): The QoS-PSO routing protocol was proposed by Liu et al. (2012) and is an agent-based routing protocol where agents are used for establishment and maintenance of routing paths. This protocol uses particle swarm optimization (PSO) for optimizing synthetic QoS. QoS can be improved by means of metrics like delay, bandwidth, packet loss, and energy.

This protocol uses forward agents and reverse agents for finding paths, where forward agents find new routes to the destination, and the case of forward agents being unable to provide QoS, then reverse agents are created and all the information from the forward agents is copied to reverse agents. Reverse agents flow in reverse direction and update the routing table at all the nodes with the new path. Iteration formulae are used for finding the new best route path, and the QoS value of this new path is calculated, and the iteration process repeats until the best path with the highest QoS value is found.

A QoS-aware routing algorithm based on ant-cluster (ICACR): ICACR was proposed by Huang et al. (2014) and is an ant-based hierarchical and multipath routing protocol which uses the concept of divide and conquer for improving the scalability of the network. This protocol works by formation of clusters, where the total number of subnetworks in the original network is one more than the total number of clusters. The additional subnetwork contains sink nodes and all of the CHs. This protocol make use of IPACR routing for finding the local optimal paths.

Feedback-enhanced learning approach (FROMS): This protocol was proposed by Förster and Murphy (2007) and is based on reinforcement learning and using local information and optimizing the route using a feedback mechanism. FROMS is an intelligent routing in-network protocol which learns about network topology by means of local information. This protocol considers equal cost path to minimize energy consumption.

This protocol is executed in three phases: sink announcement, exploration, and stable data gathering. In sink announcement, each sink broadcasts a request message to the entire network, and this broadcast message contains time to live (TTL), sink ID, and the number of hops. This message goes to each node. If the message detail is already there in their routing table but with a lesser number of hops, then the message is not processed. If the entry of this node is not in the table, then it will be updated or if the hop count is less than the entry already in the table, then the routing table is updated. In the exploration phase, the best possible route is found. The best route can be discovered by using an algorithm with route length estimate (RLE) as the metric to decide the fitness of the path equal to the number of hops. The stable data gathering phase starts when the sender node starts transmission to the sinks. This protocol minimizes energy consumption by considering equal cost paths and switching between them to increase the overall lifetime of the network

Strength pareto evolutionary algorithm (SPEA): The SPEA protocol was proposed by Kotecha and Popat (2007) and uses a genetic algorithm (GA) for optimization of various QoS parameters like bandwidth, traffic, hop-count, and delay. This protocol is adaptable for any changes within the network topology. Route discovery takes place in the MAC layer in this protocol. There are two modules for this protocol: QoS routing module and GA module. In the QoS routing module, when a data transfer request is received, it checks for the route to the destination. If this route is not available, then a route discovery procedure is initiated by transmission of smart agents. Source nodes on receiving RREP message follow reverse path and update the routing table. When an RERR message is received by the source node, it uses a GA module. Available entries in the buffer are used as initial population, and the fitness function is applied and finds the destination node with the desired tQoS requirements. If there is no RERR, then the buffer is used to find the route to destination. The buffer is evaluated periodically to maintain the efficiency of the system.

Bee sensor: This protocol was proposed by Saleem et al. (2012) and is an intelligent protocol based on honey bee optimization. This is an efficient, scalable, and energy-aware protocol. This protocol minimizes processing and communication cost.

There are four types of agents in this protocol used for setting up communication within the network as packers, scouts, foragers, and swarms. Packers are like food storer bees in a hive, receiving packets from the upper layer and finding the appropriate forager for them. Scouts search potential sink nodes by exploring the entire network. The forager uses a predetermined path for carry packets to the sink node. They also evaluate the quality of the path and transfer that information to other foragers. These foragers wait for a predetermined time at the sink node and build into swarms of waiting foragers that are transported back to the source node. Thus, this algorithm can be categorized into four parts as scouting, foraging, swarming and routing loops, and path maintenance.

EQR-RL: This protocol was proposed by Jafarzadeh and Moghaddam (2014) and is an intelligent, dynamic routing algorithm based on reinforcement learning. For updating the latest changes within a network, it uses a controlled flooding algorithm. Each node uses the header of its neighbor's data packet for getting the latest information on its neighbors. This protocol focuses on QoS requirements like latency hop-count and bandwidth.

FTIEE: FTIEE was proposed by Kiani et al. (2015) and uses Q-learning for selecting the optimal CH node and reduces the transmission cost of CH-sink node to its minimum. In this protocol each sensor node works as an independent entity as the learning agent which can choose its neighbor for forwarding data or select itself as the CH node.

Network-Structure-Based Routing Protocols

These routing protocols are based on network structure and can be classified into three subgroups as flat routing, hierarchical routing, and location based routing.

Flat routing: This is a data-centric protocol where each node has the same responsibility. In this routing the sensor node senses the environmental variables and transfers that data to the base station (BS) that is responsible for sending requests and queries. These routing protocols use data aggregation for reducing redundant data and thus reducing energy. There are two types of flat routing protocol: real-time and energy-aware routing (REAR) and load-based energy-aware multimedia routing (LEAR).

Real-time and energy-aware routing (REAR): REAR was proposed by Lan et al. (2008) and is a metadata-based routing protocol supporting query-based services. This protocol focuses on end-to-end delay and energy efficiency of the network. It make use of an optimized Dijkstra algorithm which is a modified Dijkstra algorithm for finding the routes with the lowest weight path. It makes use of metadata for the routing discovery process. QoS depends on factors like bandwidth, delay, and remaining energy. For reducing queuing delay, it make use of a "classified queuing model" which also reduces real-time transmission of data.

Load-based energy-aware multimedia routing (LEAR): LEAR was proposed by Nayyar et al. (2011b) and is a self-organizing, reactive, multipath and event driven protocol that focuses on parameters like reliability, efficiency, and load balancing. It selects the route based on "route selection factor-β" which is calculated by the forwarding nodes. There are also "swap nodes" which are nodes powered at 25% of their total power, and these nodes do not take part in routing as they are used in certain critical conditions only. For transmission of multimedia traffic, a disjoint path is selected and in the absence of a disjoint path, a partial disjoint path is selected which may increase the hop-count but ensure high-throughput, low latency, and reduce congestion which is needed for multimedia data. This protocol also efficiently avoids holes by using swap nodes.

Hierarchical Routing

Hierarchical routing protocol divides the entire network into different levels in which nodes have different responsibilities. Normally, this routing protocol consists of two routing layers and data flows from the lower layer to the higher layer. Here different clusters form within the node having maximum capabilities and responsibilities as the cluster head (CH), and the rest of the nodes as child node/cluster members. These child node/cluster members can only communicate with the CH, but the CH can communicate with the base station (BS), thus communication flows in a hierarchical manner. These protocols are better than flat routing protocol in terms of energy efficiency, reliability, and scalability. The performance of these protocols depends upon the process cluster formation. There are two protocols associated with this: energy efficient and perceived QoS-aware video routing (PEMuR) and energy efficient QoS assurance routing (EEQAR).

Energy efficient and perceived QoS-aware video routing (PEMuR): PEMuR was proposed by Kandris et al. (2011) and is an energy-aware hierarchical routing protocol using an intelligent video scheduling scheme for detecting any distortion within video. This protocol is based on the scalable power efficient routing (SHPER) protocol. This protocol handles low bandwidth situations very intelligently. In this protocol CHs are categorized into upper level CHs and lower level CHs, where upper level CHs are those that are nearer to the BS and make direct transmissions to the BS. Lower level CHs are farther from the BS, and they send data to the BS through multi hop routing.

This protocols is divided into two phases: initialization phase and steady state phase. The initialization phase creates a TDMA schedule for all of the nodes to advertise themselves and calculates the relative distances between them. In the steady state phase, the BS selects the node with maximum energy in each cluster as the CH.

This protocol reduces the retransmission rate by dropping video packets that are unable to manage availability of sufficient bandwidth. These dropped packets are chosen wisely for minimizing overall video distortion using a video distortion model. Thus, this protocol intelligently handles low bandwidth situations.

Energy efficiency QoS assurance routing (EEQAR): EEQAR was proposed by Lin et al. (2011) and is an energy efficient hierarchical routing protocol. It uses cellular topology for formation of clusters. The QoS thrust model of EEQAR is based on social network analysis, where a relationship between two individuals is determined by the trust relationship between them. This trust relationship is built using distributed metrics. Trust values are calculated using direct behavior monitoring and indirect information gathering.

Various attributes which are required for calculating trust values are frame rate, transmission delay, image quality, and audio quality. Cellular topology is created based on the position of agent and multimedia sensor nodes. After deployment, agents are used to send advertising messages with their IDs, and multimedia sensor nodes join with agent nodes that have maximum signal strength. If there is no work then multimedia sensor nodes can sleep to save energy, but agents must be active throughout the process.

There is an optimization factor table which is maintained by each node which stores parameters like trust value, energy level, and correlation of each neighbor node. This table estimates the correlation between two nodes and thus directly affects the fusion process. This factor table is used to find the next forwarding node.

This protocol works in several rounds, where each round is associated with three phases: cluster building, routing probe, and steady state. The cluster building phase deals with building the cellular cluster. The routing probe phase deals with building intercluster and intracluster routes. Steady state deals with the collection of data.

Location-Based Routing

The location-based routing protocol route packets based on exploiting knowledge of the location of nodes in the network. The location-based routing protocol uses the global positioning system (GPS) or other forms of localization techniques for determining the location of the nodes.

The location-based routing protocol can be used for minimizing routing overhead, utilizing bandwidth in an efficient manner, scalability, and tolerance on a faulty route. The location-based routing protocol is fully distributed and stateless, requiring less memory and maintenance.

The location-based routing protocol suffers from the problem of "energy hole". If the destination is in the same region as the sender, then this protocol suffers from repetitive usage of same path. There are various protocols under location-based routing protocols: two-phase geographic greedy forwarding (TPGF), geographic energy-aware multipath stream-based (GEAMS), greedy perimeter stateless routing (GPSR), directional geographic routing (DGR), energy-aware TPGF (EA-TPGF), and pairwise directional geographical routing (PWDGR).

Two-phase geographic greedy forwarding (TPGF): This protocol was proposed by Shu et al. (2010) and focuses on minimizing the "local minimum problem". It is a location-based routing protocol which supports event driven services. It is a multipath and implements a greedy forwarding (GF) algorithm. It eliminates path circles and optimizes the routing path by a using label-based optimization method. For bypassing holes, two approaches are used:

1. Advance computing of a planar graph without knowing holes or boundary nodes
2. Bypassing holes by having prior information of holes or boundary nodes

It uses a minimum number of hop-counts for finding the best path which requires prior information of the entire network and therefore does not support scalability.

Geographic energy-aware multipath stream-based (GEAMS): GEAMS is a geographical multipath routing protocol that focuses on load balancing and minimum energy consumption. It was proposed by Medjiah et al. (2009). It enforces uniform energy consumption and meets delay requirements. This protocol makes routing decisions without knowledge of global topology.

Each nodes stores information of its one-hop neighbor which is updated by means of beacon messages. This information includes estimated distance to its neighbor, distance of neighbor to sink, remaining energy, and data rate of link. This protocol works on two policies: smart greedy forwarding and walking back forwarding. Walking back forwarding is used when no neighbor is closer to the sink, otherwise smart greedy forwarding is used.

Greedy perimeter stateless routing (GPSR): This is a location-based routing protocol which can be used for detection and avoidance of hole problems. This protocol was proposed by Karp and Kung (2000) and supports MAC-layer failure feedback. This protocol makes use of the position of routers and destination node for making decisions on packet forwarding. It uses greedy forwarding and perimeter forwarding for packets within the network. GPSR makes use of flags for indicating the current forwarding mode of packets. GPSR uses perimeter forwarding by forwarding packets based on planar graph traversal. If a packet reaches a region where greedy forwarding is not possible, GPSR recovers by routing the packet around the perimeter of the region. It uses a promiscuous mode and supports queue traversal, implementing planarization of graphs.

Directional geographic routing (DGR): This protocol was proposed by Chen et al. (2007) and is a location-based protocol which uses disjoint paths for optimizing energy, bandwidth, and delays. DGR is a fault tolerant routing protocol which supports load balancing, fast packet delivery, and good peak signal-to-noise ratio (PSNR). DGR gives better quality video transmission and prolongs network lifetime.

This protocol assumes that any node can send a video packet to the sink at any instance. It works by forming multiple application-specific disjoint paths for transmitting real-time traffic over low bandwidth on an unreliable network. It uses forward error correction (FEC) coding for reliable delivery of data. This protocol divides a single real-time video stream into multiple substreams that are transferred in parallel through disjoint paths for full utilization of resources.

Energy-aware TPGF (EA-TPGF): This is an enhanced version of the TPGA protocol which was proposed by Bennis et al. (2013). This protocol focuses on energy efficiency and uses a distance-energy formula and energy cost transmission formula for evaluating a path for the route. Like the TPGF protocol, it uses a greedy forwarding (GF) algorithm for routing data packets. It uses step back and mark method for situations with no single node in the neighborhood closer to the BS. It also deals with label-based optimizations of the path. It uses a distance-energy formula for calculating the score of neighbors where the node with minimum value is selected. It also defines the energy cost transmission formula for calculating the cost of energy consumed by the transmitter.

Pairwise directional geographical routing (PWDGR): This is a multipath-based location routing protocol that uses the concept of pairwise nodes to fully utilize the nodes around the sink. This protocol was proposed by Wang et al. (2015) and increases the efficiency of network lifetime by 70%. It is a reliable and fault tolerant protocol which improves bandwidth on the network.

Routing in this protocol is carried out in three stages. The first stage is collection and packaging of data that is then broadcasted, and this saves energy and reduces delay. The second stage includes receiving of neighbor nodes, and checks to see if they are in the cooperative node list. If yes, then the second stage is calculation of the coordinates of the next neighbor node. The third and final stage selects a source cooperative node based on two rules: (1) cumulative numbers to be selected for cooperative node; and (2) distance between the ideal point and the neighbor node.

References

A. A. Ahmed and N. Fisal, A real-time routing protocol with load distribution in wireless sensor networks. *Computer Communications*, vol. 31, no. 14, pp. 3190–3203, 2008.

K. Akkaya and M. Younis, An energy-aware QoS routing protocol for wireless sensor networks. In: *Proceedings of the 23rd International Conference on Distributed Computing Systems Workshops (ICDCSW '03)*, 2003, pp. 710–715.

I. F. Akyildiz, T. Melodia and K. R. Chowdhury, A survey on wireless multimedia sensor networks. *Computer Networks*, vol. 51, no. 4, pp. 921–960, 2007.

I. Bennis, O. Zytoune, D. Aboutajdine and H. Fouchal, Low energy geographical routing protocol for wireless multimedia sensor networks. In: *Proceedings of the 9th International Wireless Communications and Mobile Computing Conference (IWCMC '13)*, 2013, pp. 585–589.

J. Ben-Othman and B. Yahya, Energy efficient and QoS based routing protocol for wireless sensor networks. *Journal of Parallel and Distributed Computing*, vol. 70, no. 8, pp. 849–857, 2010.

J. Bi, Z. Li and R. Wang, An ant colony optimization-based load balancing routing algorithm for wireless multimedia sensor networks. In: *Proceedings of the IEEE 12th International Conference on Communication Technology (ICCT '10)*, 2010, pp. 584–587.

M. Chen, V. C. M. Leung, S. Mao and Y. Yuan, Directional geographical routing for real-time video communications in wireless sensor networks. *Computer Communications*, vol. 30, no. 17, pp. 3368–3383, 2007.

O. Chipara, Z. He, G. Xing and Q. Chen, Real-time power-aware routing in sensor networks. In: *Proceedings of the 14th IEEE International Workshop on Quality of Service (IWQoS '06)*, 2006, pp. 83–92.

L. Cobo, A. Quintero and S. Pierre, Ant-based routing for wireless multimedia sensor networks using multiple QoS metrics. *Computer Networks*, vol. 54, no. 17, pp. 2991–3010, 2010.

G. Di Caro and M. Dorigo, AntNet: Distributed stigmergetic control for communications networks. *Journal of Artificial Intelligence Research*, vol. 9, pp. 317–365, 1998.

S. Ehsan and B. Hamdaoui, A survey on energy-efficient routing techniques with QoS assurances for wireless multimedia sensor networks. *IEEE Communications Surveys and Tutorials*, vol. 14, no. 2, pp. 265–278, 2012.

E. Felemban, C.-G. Lee and E. Ekici, MMSPEED: multipath multi-SPEED protocol for QoS guarantee of reliability and timeliness in wireless sensor networks, *IEEE Transactions on Mobile Computing*, vol. 5, no. 6, pp. 738–753, 2006.

A. Förster and A. L. Murphy, FROMS: Feedback routing for optimizing multiple sinks in WSN with reinforcement learning. In: *Proceedings of the 3rd International Conference on Intelligent Sensors, Sensor Networks and Information (ISSNIP '07)*, IEEE, 2007, pp. 371–376.

T. He, J. A. Stankovic, C. Lu and T. Abdelzaher, SPEED: A stateless protocol for real-time communication in sensor networks. In: *Proceedings of the 23rd IEEE International Conference on Distributed Computing Systems*, 2003, pp. 46–55.

H. Huang, X. Cao, R. Wang and Y. Wen, A QoS-aware routing algorithm based on ant-cluster in wireless multimedia sensor networks. *Science China Information Sciences*, vol. 57, no. 10, pp. 1–16, 2014.

S. Z. Jafarzadeh and M. H. Y. Moghaddam, Design of energy-aware QoS routing protocol in wireless sensor networks using reinforcement learning. In: *Proceedings of the IEEE 27th Canadian Conference on Electrical and Computer Engineering (CCECE '14)*, 2014, pp. 1–5.

D. Kandris, M. Tsagkaropoulos, I. Politis, A. Tzes and S. Kotsopoulos, Energy efficient and perceived QoS aware video routing over wireless multimedia sensor networks, *Ad Hoc Networks*, vol. 9, no. 4, pp. 591–607, 2011.

B. Karp and H. T. Kung, GPSR: Greedy perimeter stateless routing for wireless networks. In: *Proceedings of the 6th Annual International Conference on Mobile Computing and Networking (MOBICOM '00)*, 2000, pp. 243–254.

F. Kiani, E. Amiri, M. Zamani, T. Khodadadi and A. A. Manaf, Efficient intelligent energy routing protocol in wireless sensor networks. *International Journal of Distributed Sensor Networks*, vol. 2015, pp. 13, 2015.

K. Kotecha and S. Popat, Multi objective genetic algorithm based adaptive QoS routing in MANET. In: *Proceedings of the IEEE Congress on Evolutionary Computation (CEC '07)*, 2007, pp. 1423–1428.

S. Kumar, M. Dave and S. Dahiya, ACO based QoS aware routing for wireless sensor networks with heterogeneous nodes. In: *Emerging Trends in Computing and Communication, vol. 298 of Lecture Notes in Electrical Engineering*. Springer: Berlin, Germany, 2014, pp. 157–168.

Y. Lan, W. Wenjing and G. Fuxiang, A real-time and energy aware QoS routing protocol for multimedia wireless sensor networks. In: *Proceedings of the 7th World Congress on Intelligent Control and Automation (WCICA '08)*, 2008, pp. 3304–3309.

B. Li and K.-I. Kim, A novel real-time scheme for (m, k)-firm streams in wireless sensor networks, *Wireless Networks*, vol. 20, no. 4, pp. 719–731, 2014.

K. Lin, J. J. P. C. Rodrigues, H. Ge, N. Xiong and X. Liang, Energy efficiency QoS assurance routing in wireless multimedia sensor networks. *IEEE Systems Journal*, vol. 5, no. 4, pp. 495–505, 2011.

M. Liu, S. Xu and S. Sun, An agent-assisted QoS-based routing algorithm for wireless sensor networks, *Journal of Network and Computer Applications*, vol. 35, no. 1, pp. 29–36, 2012.

C. Lu, B. M. Blum, T. F. Abdelzaher, J. A. Stankovic and T. He, RAP: A real-time communication architecture for large-scale wireless sensor networks. In: *Proceedings of the 8th IEEE Real-Time and Embedded Technology and Applications Symposium (RTAS '02)*, 2002, pp. 55–66.

S. Medjiah, T. Ahmed and F. Krief, GEAMS: A geographic energy-aware multipath stream-based routing protocol for WMSNs. In: *Prossceedings of the Global Information Infrastructure Symposium (GIIS '09)*, 2009, pp. 1–8.

S. Medjiah, T. Ahmed and F. Krief, AGEM: Adaptive greedy-compass energy-aware multipath routing protocol for WMSNs. In: *Proceedings of the 7th IEEE Consumer Communications and Networking Conference (CCNC '10)*, 2010, pp. 1–6.

A. Nayyar, F. Bashir and Ubaid-Ur-Rehman, Load based energy aware multimedia routing protocol-(LEAR). In: *Proceedings of the 3rd International Conference on Computer Research and Development (ICCRD '11)*, vol. 2, 2011a, pp. 427–430.

A. Nayyar, F. Bashir, Ubaid-Ur-Rehman and Z. Hamid, Intelligent routing protocol for multimedia sensor networks. In: *Proceedings of the 5th International Conference on Information Technology and Multimedia (ICIM '11)*, 2011b, pp. 1–6.

A. Rahman, R. Ghasemaghaeil, A. El Saddikl and W. Gueaieb, M-IAR: Biologically inspired routing protocol for wireless multimedia sensor networks. In: *Proceedings of the Instrumentation and Measurement Technology Conference (IMTC '08)*, 2008, pp. 1823–1827.

Y. Rao, C. Yuan, Z. Jiang, L. Fu and J. Zhu, Latency and reliability-aware geographic routing for mobile wireless sensor networks. *Advances in information Sciences and Service Sciences*, vol. 5, no. 8, pp. 738–748, 2013.

S. Ratnaraj, S. Jagannathan and V. Rao, OEDSR: Optimized energy-delay sub-network routing in wireless sensor network. In: *Proceedings of the IEEE International Conference on Networking, Sensing and Control (ICNSC '06)*, 2006, pp. 330–335.

M. A. Razzaque, M. M. Alam, M. Mamun-Or-rashid and C. S. Hong, Multi-constrained QoS geographic routing for heterogeneous traffic in sensor networks. *IEICE Transactions on Communications*, vol. 91, no. 8, pp. 2589–2601, 2008.

M. Saleem, I. Ullah and M. Farooq, BeeSensor: An energy-efficient and scalable routing protocol for wireless sensor networks. *Information Sciences*, vol. 200, pp. 38–56, 2012.

H. Shen, G. Bai, Z. Tang and L. Zhao, QMOR: QoS-aware multi-sink opportunistic routing for wireless multimedia sensor networks. *Wireless Personal Communications*, vol. 75, no. 2, pp. 1307–1330, 2014.

L. Shu, Y. Zhang, L. T. Yang, Y. Wang, M. Hauswirth and N. Xiong, TPGF: Geographic routing in wireless multimedia sensor networks. *Telecommunication Systems*, vol. 44, no. 1–2, pp. 79–95, 2010.

K. Sohrabi, J. Gao, V. Ailawadhi and G. J. Pottie, Protocols for self-organization of a wireless sensor network. *IEEE Personal Communications*, vol. 7, no. 5, pp. 16–27, 2000.

Y. Sun, H. Ma, L. Liu and Y. Zheng, ASAR: An ant-based service-aware routing algorithm for multimedia sensor networks. *Frontiers of Electrical and Electronic Engineering in China*, vol. 3, no. 1, pp. 25–33, 2008.

J. Wang, Y. Zhang, J. Wang, Y. Ma and M. Chen, PWDGR: pair-wise directional geographical routing based on wireless sensor network. *IEEE Internet of Things Journal*, vol. 2, no. 1, pp. 14–22, 2015.

Y. Xue, B. Ramamurthy and M. C. Vuran, SDRCS: A service-differentiated real-time communication scheme for event sensing in wireless sensor networks. *Computer Networks*, vol. 55, no. 15, pp. 3287–3302, 2011.

X. Yu, J. Luo and J. Huang, An ant colony optimization-based QoS routing algorithm for wireless multimedia sensor networks. In: *Proceedings of the 3rd International Conference on Communication Software and Networks (ICCSN '11)*, 2011, pp. 37–41.

L. Zhou and J.-X. Wang, Delay-constrained maximized lifetime routing algorithm in wireless multimedia sensor networks. In: *Proceedings of the 2nd International Conference on Future Computer and Communication (ICFCC'10)*, 2010, pp. V1215–V1219.

Chapter 7

Clustering in Wireless Sensor Networks

A wireless sensor network (WSN) is a network which consists of a number of tiny nodes which are equipped with the ability to communicate with one another. A node in the network can easily be seen as a very compact packaging of multiple units viz. sensing unit, processing unit, transceiving unit, power unit, and an optional mobility and power supply unit (if needed by the concerned applications). All such nodes are deployed randomly in a sensing field which is usually inaccessible to humans and in harsh climate conditions for most of the applications. This simply means that the once nodes are deployed, they cannot be altered. This particular deployment strategy requires the network to be self-organizing and autonomous where self-organization refers to the ability of the nodes to network without being in a fixed infrastructure, and autonomy means sensor nodes must be able to decide what is required to meet the needs of the application in an efficient way.

Wireless sensor networks are resource constrained in nature, especially because of the tiny size of the constituent nodes. The small node size imposes a lot of restrictions over the network, among which a limited *power restriction* is the most important one to consider. Also, the network is comprised of a huge number of nodes which continue to increase, and this necessitates provisioning for a *scalable*, power efficient network, especially with regard to transmission, as it this has been proved to be the most energy consuming task for a node in the network. Thus, it can be concluded that energy-efficiency and scalability are the two basic requirements for any scheme to be employed in WSNs.

To meet the above cited objectives, energy efficiency and scalability, the notion of hierarchical architecture came into existence, in which the network is divided into a number of layers; each layer holding the nodes of peer functionality, i.e., the nodes lying in the same layer play the same role. In wireless sensor networks, the idea of hierarchical architecture is implemented as 2-tier clustering, where clustering is defined as the grouping of similar objects or the process of identifying the natural association among the objects (Ghiasi et al., 2002). In order to achieve energy efficiency and scalability, the majority of the nodes are partitioned into a number of clusters at lower level to pursue their intended activities of sensing and transmitting the sensed data to the smaller number of the specially designated nodes called cluster heads which perform more complex tasks viz. communicating with the base station. Moreover, energy is also conserved in this approach via data aggregation. Since the large volume of data being produced in the network is spatiotemporally correlated, data aggregation is implemented by the cluster heads in their respective clusters to reduce the size of transmittable data, as a larger size data packet means higher energy consumption in transmission. Cluster heads receive data from their respective cluster members, aggregate it, and then transmit the processed data to the base station either using a single hop or multi hop transmission approach as provisioned. The end user can access the data at this point through some appropriate internet connectivity and utilize it in the respective application. The concept of clustering is elaborated in Figure 7.1.

Different nodes are grouped together on the basis of some common attributes into a number of clusters and are referred to as cluster members. Each cluster is represented by specially designated nodes called s cluster head which collects and processes the data received from the respective cluster members. The data within the clusters is referred to as intra-cluster traffic. Moreover, the data from the clusters is transported to the base station either using a direct or multi hop approach. The data between the different clusters is referred to as inter-cluster traffic. Since the cluster head is the node which performs comparatively more energy consuming tasks viz. collecting and aggregating the data received from its members and

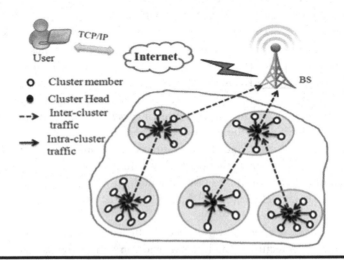

Figure 7.1 Clustering architecture in WSNs.

transmitting such processed data to the base station, the battery of the cluster head may deplete more rapidly. Therefore, in order to achieve load balancing among the nodes in network, the cluster heads are rotated.

Since its conception, clustering has proved to be an efficient scheme especially when employed in routing, not only contributing to energy efficiency, but also a number of other objectives viz. improved coverage, enhanced network lifetime, and fault tolerance can be achieved. The scheme is basically divided into three phases:

1. Set up
2. Selection
3. Data transmission

In the setup phase, the network is divided into a number of clusters. Each cluster is assigned a cluster head. In the selection phase, cluster heads are chosen/decided. However, in some clustering-based schemes, the set up and selection phase are combined to form a single phase. In the next phase of data transmission, sensor data is sent to the respective cluster head from where it is transmitted to the base station.

This chapter highlights the various objectives to be achieved using clustering. Furthermore, there is detailed discussion of the prominent clustering characteristics. Finally, the chapter concludes by reviewing the popular clustering-based routing protocols in subsequent sections.

Clustering Objectives

Wireless sensor networks are networks which are very specific to the applications being deployed and, obviously, every application has its own requirements defining the objective or purpose of network. Clustering has proved itself to be an important tool in achieving these objectives and purposes. Such objectives are classified into two major categories: explicit/direct and implicit/indirect objectives. Explicit objectives are the ones which are the obvious results of the clustering process and can be determined substantially, whereas implicit ones are those which are obtained indirectly as the result of the clustering process. Figure 7.2 lists and discuss the major objectives which are achieved as outcomes of clustering.

Explicit or Direct Objectives

Scalability

Scalability refers to the solution being independent of the size of the network, i.e., irrespective of the node population in the network, the given solution must work. Since the number of nodes deployed in the sensing field depends heavily on the nature of the application and may go up to an order of hundreds, thousands, or even millions (Akyildiz et al., 2002). Clustering accommodates all such deployed

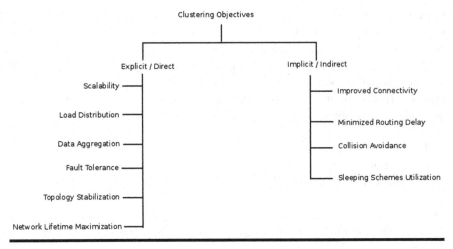

Figure 7.2 Classification of clustering objectives.

nodes into a number of clusters where if a node in a cluster, say "A" wants to communicate with some node in a different cluster, say "B", it must get equipped with the knowledge of the respective cluster head of "B". In this way, scalability is increased without any significant increase in routing table entries.

Load Distribution

As per the intrinsic formulation of clusters, nodes are divided into two (or more) tiers where nodes on a particular level share a common responsibility. More illustratively, nodes in the lower level are meant to send their environment measures to their respective cluster heads only, whereas cluster heads situated in the higher level are responsible for long range communication (to the base station), data collection (from the member nodes), and data aggregation. Thus, it can be easily intuited that cluster heads suffer from early depletion of their energy. Therefore, in order to assure even distribution of the network load, roles assigned to the nodes are changed periodically.

Moreover, when compared to a flat architecture, in a clustered network, only a few nodes are required to transmit to the base station; therefore, the major cause of energy consumption is limited to a few nodes.

Data Aggregation

Due to the close proximity among the nodes deployed in the sensing field, data generated from the network has a higher degree of spatial and temporal correlation. If a flat network architecture is implemented, base station may receive a huge volume of redundant data. Clustering helps in this repect by provisioning data aggregation at cluster level so that a normalized data set is prepared and forwarded to the base station from each of the clusters.

Fault Tolerance

Fault tolerance refers to the property in which the network succeeds in achieving the specified goals and functionality, even in the presence of faults. Clustering allows some degree of fault tolerance due to its intrinsic nature of grouping the nodes. As the nodes closer to one another are grouped together, and data generated by these nodes are quite similar, the failure of a few nodes in the cluster will not affect network performance. Even faults at a higher level, i.e., among the cluster heads, can be handled using adaptive clustering (Heinzelman et al., 2000) by provisioning for reclustering of nodes at the beginning of predetermined rounds. Moreover, other than costlier reclustering, appointing backup cluster heads or associate cluster heads is also a popular solution for achieving fault tolerance by implementing clustering of nodes.

Topology Stabilization

Since the clustering provisions nodes to be grouped together, any intra-cluster movements of the nodes can be managed at cluster level itself. As per the clustering process, since cluster heads are aware of their respective members' current location and other important information viz. residual energy etc., any movement or events like a node's death can be reported by the respective cluster head. Therefore, contrary to a flat architecture with/without mobile nodes, network topology is more stable in a clustered wireless sensor network.

Network Lifetime Maximization

Network lifetime describes the time for which network remains operating, i.e., the nodes deployed can sense and report to the base station. It has already been mentioned that the ultimate objective behind the notion of clustering is energy efficiency. Energy efficiency achieved through aggregation, even load distribution, etc. results in network lifetime maximization. The lifetime of a network can also be further maximized through additional policies like situating cluster heads comparatively at central position in their respective clusters (so that members spend equal energy in communicating to their cluster heads), and utilizing the sleeping nodes strategy.

Implicit or Indirect Objectives

Improved Connectivity

Clustering results in a *k-connected network* where k is the number of members in a cluster implying that it requires *k node failures* to leave the network disconnected. When compared to flat architecture, where almost every node is required to be connected to base station, in a clustered network, it is required to assure paths only between the cluster heads and base station.

Minimized Routing Delay

In a flat network architecture, generally the multi hop approach is followed for communicating the network measures to the base station and the length of the path is usually of the order of n where n is the number of nodes in the network that definitely results in a huge delay, especially in the context of time critical applications viz. multimedia and health monitoring applications. When the flat architecture is replaced by clustering architecture, even in the worst case, the path length is reduced to k, where k is the number of clusters formed in the network as the sensed data may go through k cluster heads at most, and hence the routing delay is minimized in clustering.

Collision Avoidance

Collision among the packets by different nodes in network is a major problem in wireless sensor networks, as it may result in a huge wastage of energy due to required retransmission in order to satisfy the needs of the deployed network. When nodes are clustered into groups, such collisions may be avoided by employing one of the available collision avoidance mechanism viz. time division multiple access (TDMA) or carrier sense multiple access (CSMA).

Sleeping Schemes Utilization

Sleeping schemes are very useful tools in order to save energy in the nodes, as instead of keeping each node awake, activating only a few nodes at a time might save overall network energy, as when nodes are brought into sleep mode, their radios are off, and they are in low power mode where they spend energy in the order of microWatt (μW) instead of milliWatt (mW) as in active mode. In fact, there are a large number of applications in which it is not required to keep every node in an activated state, so the nodes are provisioned to go into sleep mode periodically. In other applications, nodes with higher coverage may be activated while putting the nodes under themselves into sleep mode, and overall energy consumption is reduced (Ye et al., 2003). However, implementing a sleeping strategy is quite simple in cluster-based networks via application of the TDMA schedule. Cluster heads can easily define the sleeping schedule for their respective cluster members and propagate the schedule into the respective clusters. Members in the cluster tune themselves as per the propagated schedule and save energy.

Clustering Characteristics

As discussed above, clustering consists of three main phases: setup, selection, and data transmission. Clusters take shape in the setup phase, cluster heads are chosen/decided in the selection phase, and data transmission takes place in the data transmission phase. Each of the above-mentioned phases contributes in clustering characteristics like cluster-specific, specific to cluster heads (CHs) and, lastly, specific to the cluster formation process (Afsar et al., 2014). Each of such characteristics would be discussed in detail in following subsections. Moreover, all these can also be summarized as in the following Figure 7.3.

Cluster-Specific Characteristics

Characteristics defined on the basis of the number of clusters formed in the network, number of cluster members, intra-cluster, and inter cluster communication are referred to as cluster-specific characteristics.

Size: refers to the number of nodes which belong to a particular cluster. Cluster size is defined on the basis of load distribution among the nodes in the network, and clusters may have an equal or unequal number of nodes in themselves after their formation. The distance between the nodes and the base station also plays a very significant role in the derivation of cluster size.

Count: number of clusters formed in the network may vary with time or may be even fixed depending upon the nature of the clustering algorithm. In the algorithms viz. (Heinzelman et al., 2000, 2002) where cluster heads are elected randomly in every round, the count may be variable; whereas, in the algorithms viz. (Zahmati et al., 2007; Chaurasiya et al., 2011) where the number of clusters was defined by the base station and does not change with time, the count may be static (constant).

Intra-cluster communication: refers to communication among the nodes within the cluster. Depending upon the clustering scheme, the approach may be single hop between the cluster head (CH) and its members, if the cluster size is small, as the distance between the CH and the members is not far. However, if the scheme results in large size clusters, where distance between a CH and its members is quite large, the approach to be followed for intra-cluster communication may be multi hop.

Inter-cluster communication: Contrary to intra-cluster communication, inter-cluster communication describes the message exchange among the cluster heads in the network. Communication may follow the multi hop approach, if the cluster count is high and the distance between a cluster head and the base station is beyond the short-range communication ability. However, the single hop approach is the usual one and is followed when the distance between the cluster head and the sink is not that much and within their short range communication ability.

CH-specific characteristics: since cluster heads have the most energy consuming tasks in the network, their characteristics viz. whether they are mobile or immobile, their ability, and their role in the network are crucial.

Mobility: in some of the applications, cluster heads are required to be mobile where they can move up to a limited distance; however, in such application topology management becomes a very challenging task with respect to the applications requiring non-mobile, i.e., stationary cluster heads.

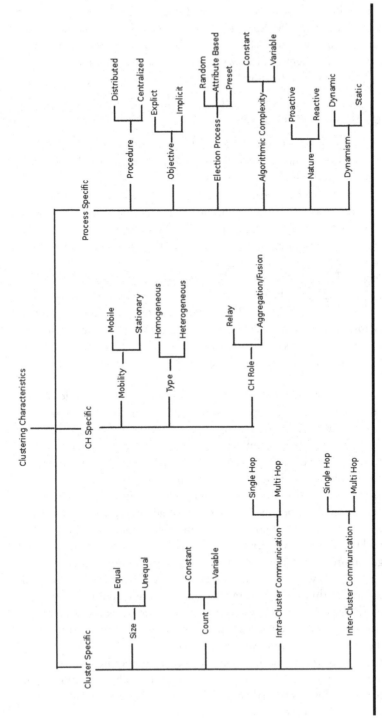

Figure 7.3 Clustering characteristics (Redrawn from Afsar et al., 2014).

Type: based on the network being deployed, nodes may be normal or advanced or even super nodes (i.e., homogeneous, 2-level heterogeneous, and 3-level heterogeneous network). If the cluster heads being chosen are rich in resources, they are known as heterogeneous, or they are homogeneous, if they are being chosen from among nodes with similar abilities.

CH role: based on the application requirement, cluster heads may perform data aggregation/fusion or simply act as relay nodes forwarding local traffic to the base station directly (single hop communication) or to some other cluster head towards the base station (multi hop communication).

Process-specific characteristics: these characteristics are specific to the process of cluster formation, i.e., to the clustering algorithm as follows:

Procedure: clustering algorithms result in the procedures for cluster formation which may be categorized as distributed or centralized. However, due to the intrinsic high node population of wireless sensor networks, distributed ones are the most popular.

Objective: defines the purpose of the network which may be categorized as explicit or implicit. Explicit objectives are the ones which are directly achievable by the network, e.g., scalability and fault tolerance, whereas, implicit objectives are the ones which refer to the indirect results of the clustering process, e.g., improved connectivity, and reduced delay. The above-mentioned characterization has already been dealt with in adequate details in the previous section of this chapter.

Election processes: for choosing the cluster heads is categorized into three: preset, random, and attribute based (Afsar et al., 2014). As implied by the name, the preset process refers to the algorithms in which cluster heads are predetermined, even before the deployment of nodes in the sensing field. Random process refers to the algorithms in which cluster heads are chosen randomly. In the attribute-based process, one or more parameters viz. residual energy, distance from base station, or neighbors associated derives the cluster heads.

Algorithmic complexity: refers to the time measure required by the algorithm to converge (Afsar et al., 2014). Algorithmic complexity can be categorized into variable time of the network specification such as the number of cluster heads, etc. are required by the algorithm to converge, and constant time if the convergence does not require that network specification.

Nature: as in the case of a normal ad hoc network, clustering algorithms can also be classified as proactive or reactive. However, reactive networks are found less often in comparison to proactive ones in sensor networks; only data centric methods implement the reactive networks. Moreover, it has been observed that some schemes use a hybrid of both proactive and reactive.

Dynamism: on the basis of the information being used to form clusters and to elect cluster heads, clustering dynamism is classified into two: dynamic and static. In the dynamic schemes, real time parameters, i.e., the current network conditions are used, whereas static dynamism uses preframed information to elect cluster heads and in other network operations irrespective of the current network condition.

Clustering Algorithms

A huge number of clustering schemes exists in the literature and have been classified into a number of ways based on different criterion as follows:

Distributed vs. centralized schemes: It can be intuited easily that due to a large population of nodes in a WSN, distributed schemes are the obvious choice, as the centralized schemes suffers with a number of limitations viz. a single point of failure, which may lead to early death of the network. Also, distributed methods are scalable which suits one of the primary requirements of WSN.

Random vs. deterministic Schemes: This classification is based on how the cluster heads are being chosen in the network. Random schemes are the schemes in which cluster heads are chosen randomly (such schemes are also referred to as probabilistic schemes). However, in deterministic schemes, cluster heads are derived schematically, and the schemes can be further categorized as weight-based, fuzzy-based, heuristic-based, and compound approach (Afsar et al., 2014).

Equal vs. unequal clustering schemes: This classification is based on how the nodes are being grouped together to form clusters in a network. Depending upon the load distribution and distance from the base station, clustering schemes yield equal or unequal sized clusters.

Other than the aforementioned classifications, there exists several other classifications to categorize the vastly available clustering schemes; however, in this section, focus will be given to classifying schemes according to the frequency with which cluster formation happen, i.e., how dynamically cluster formation takes place in the network. In view of the above, clustering schemes are divided into two classes: dynamic and static. In the dynamic clustering schemes, clusters are formed/ reformed periodically (after every round), whereas in the static clustering schemes, clusters decided once remain fixed throughout the network lifetime, until the end of network operation. Random clustering schemes under the dynamic category for both homogeneous and heterogeneous networks and weight-based deterministic clustering schemes are discussed in the following sections.

Dynamic clustering schemes: In dynamic clustering techniques, clusters are formed periodically after every round. Many clustering schemes have already been proposed for this category out of which a few are listed below in the context of both homogeneous and heterogeneous network respectively.

Low Energy Adaptive Clustering Hierarchy (LEACH): Low energy adaptive clustering hierarchy (LEACH) is the first scheme which explores the advantages of clustering at its most, i.e., LEACH is the first clustering-based routing scheme for wireless sensor networks devised by Heinzelman et al. (2000). The key features of the scheme are:

1. Localized coordination and control for cluster setup and operation
2. Randomized rotation of the cluster heads and the corresponding clusters
3. Local compression to reduce global communication (Heinzelman et al., 2000)

Localized coordination among the nodes consolidates the self-organization property of WSNs. It restricts the use of any outliers in the formation of clusters in a network, i.e., nodes are required to coordinate and cooperate with one another in sharing the information to form clusters. Randomized rotation of the role of the cluster head (CH) among the nodes in the network confirms load balancing or even load distribution, as every node is supposed to play the role of CH which consumes a lot of energy in comparison to normal nodes. Lastly, local compression of data is an obvious advantage of clustering the nodes together in network. Since nodes in the cluster are close to each other, they may produce similar if not the same data from the concerned field. Hence, data aggregation is used to yield a single packet of data from the packets received by the cluster members and that is forwarded to the base station for further access by the end user. This reduces the amount of data to be sent over the medium and conserves energy.

The network operation of LEACH is divided into two phases: set up phase and steady state phase.

In the set up phase, every node generates a random number and compares such generated number with T(n), and here T(n) is given as follows:

$$T(n) = \begin{cases} \dfrac{p}{1 - p \times \left(r \bmod \dfrac{1}{p} \right)} & \text{if } n \in G \\ 0 & \text{otherwise,} \end{cases}$$

If the randomly generated number within the node is less than T(n), the node declares itself the cluster head and invites other nodes to join him as cluster members. In the above formula, p is a suggested percentage of cluster heads and r is the current round, and G is the set of nodes in the network that have not been selected as cluster heads in the previous (1/p) rounds. Nodes other than those who have generated numbers less that T(n), i.e., other than CH candidates, join the nearest CH on the basis of received signal strength.

Once the clusters, cluster members, and cluster heads are decided, cluster heads broadcast a TDMA schedule for their corresponding members to end the set up phase.

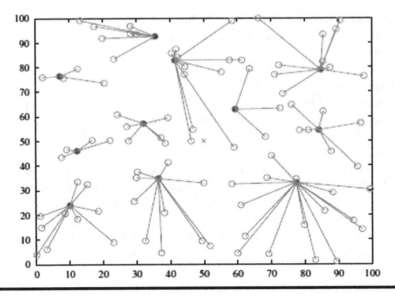

Figure 7.4 An instance of cluster formation in LEACH.

In the steady state phase, cluster members send their sensed information to the concerned CH, which in turn aggregates and forwards the data to the base station.

LEACH is a completely distributed algorithm which selects CHs probabilistically, hence, it neither guarantees the number of clusters nor how well they will be distributed in view of energy loads (Figure 7.4).

In order to compensate for these inefficiencies, LEACH-C was devised by the same authors, Heinzelman et al. (2002). In this improved scheme, each node is required to send its energy and location information to the base station, which in turn forms suitable/appropriate clusters and provisions for centralized CHs using a simulated annealing algorithm. However, LEACH-C proceeds with a very unrealistic assumption that every node would have adequate energy to communicate with the base station.

Later on, a number of schemes were introduced which either offer some refinements or an extended version of LEACH.

In the following subsection, an instance of clustering in heterogeneous wireless sensor networks is discussed.

Enhanced Developed Distributed Energy-Efficient Clustering (EDDEEC) for Heterogeneous WSNs

With the deployment of nodes with additional abilities in a homogeneous sensor network, the network becomes a heterogeneous wireless sensor network (HWSN), and it becomes quite obvious that the schemes devised for a homogenous network is not suitable in this scenario, as it may fail to handle nodes properly. Originally, a clustering-based routing scheme for heterogeneous WSN was envisioned by Qing et al. (2006) in their scheme entitled, "Design of a Distributed Energy-Efficient Clustering Algorithm for Heterogeneous Wireless Sensor Network (DEEC)". DEEC considers two-level energy heterogeneity in WSN, i.e., in such networks, two types of nodes, say normal nodes and advanced nodes, exist with different initial energies where the energy of the advanced nodes are a-times more than that of normal nodes. DEEC improves the cluster head election process proposed in LEACH (Heinzelman et al., 2000, 2002) to compete with the changed network scenario of HWSN. In DEEC, the probability used to evaluate the cluster heads is defined as the ratio between the residual energy of nodes and average residual energy of network as follows:

$$
p_i = \begin{cases} \dfrac{p_{opt} E_i\left(r\right)}{\left(1+ am\right)\overline{E}\left(r\right)} & \text{if } S_i \text{ is the normal node} \\[4mm] \dfrac{p_{opt}\left(1+ a\right)E_i\left(r\right)}{\left(1+ am\right)\overline{E}\left(r\right)} & \text{if } S_i \text{ is the advanced node} \end{cases}.
$$

where m is the fraction of total number of nodes deployed in the network (i.e., N) which are advanced in nature; E_i is the residual energy of ith node and E' is the average residual energy of network defined as below:

$$
\overline{E}\left(r\right) = \frac{1}{N}\sum_{i=1}^{N} E_i\left(r\right).
$$

Later on the scheme proceeds as in LEACH regarding cluster formation and data transfer.

As the time has gone on, two-level heterogeneity has been enhanced to three-level and above as shown in the following Figure 7.5.

In Figure 7.5, A refers to advanced nodes, N refers to normal nodes, and S refers to super nodes.

Enhanced developed distributed energy-efficient clustering for heterogeneous wireless sensor networks (EDDEEC) was proposed by Javaid et al., 2013) to deal with a three-level heterogeneous sensor network which is basically an improvement over the EDEEC (Saini and Sharma, 2010). EDDEEC accommodates three types of nodes:normal, advanced, and super nodes where advanced nodes' initial energy is a times greater than that of normal nodes, and super nodes' initial energy is b times greater than that of normal nodes. The probability formula for cluster head selection in EDDEEC is given as follows:

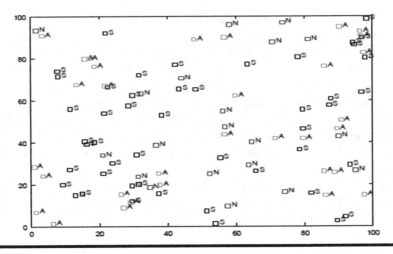

Figure 7.5   Network with 3-level energy heterogeneity.

$$
p_i = \begin{cases}
\dfrac{p_{opt} E_i(r)}{\left(1 + m\left(a + m_o b\right)\right)\bar{E}(r)} & \text{for } N_{ml} \text{ nodes} \\[2pt]
& \text{if } E_i(r) > T_{\text{absolute}} \\[8pt]
\dfrac{p_{opt}\left(1 + a\right) E_i(r)}{\left(1 + m\left(a + m_o b\right)\right)\bar{E}(r)} & \text{for } Adv \text{ nodes} \\[2pt]
& \text{if } E_i(r) > T_{\text{absolute}} \\[8pt]
\dfrac{p_{opt}\left(1 + b\right) E_i(r)}{\left(1 + m\left(a + m_o b\right)\right)\bar{E}(r)} & \text{for } Sup \text{ nodes} \\[2pt]
& \text{if } E_i(r) > T_{\text{absolute}} \\[8pt]
c\dfrac{p_{opt}\left(1 + b\right) E_i(r)}{\left(1 + m\left(a + m_o b\right)\right)\bar{E}(r)} & \text{for } N_{ml}, Adv, Sup \text{ nodes} \\[2pt]
& \text{if } E_i(r) \leq T_{\text{absolute}}
\end{cases}
$$

Here, m and m_0 are the proportion of advanced and super nodes in the network with total N number of nodes. T_{absolute} is a new constant provisioned in the scheme in order to achieve a balanced load distribution when residual energy of advanced nodes and super nodes becomes close to that of normal nodes. T_{absolute} is defined as zE_0 where z is experimental constant set to 0.7 and E_0 is the initial energy of normal nodes.

Once the cluster heads are decided, the scheme proceeds as it does in LEACH (Heinzelman et al., 2000, 2002) or DEEC (Qing et al., 2006) for cluster formation and data transfer (Figure 7.6).

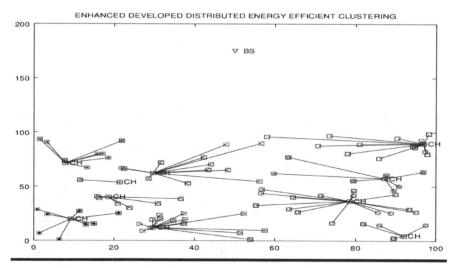

Figure 7.6   An instance of cluster formation in EDDEEC.

Static Clustering Schemes

As discussed in the previous section, energy-efficiency is the most important objective which has been pursued successfully in the above mentioned dynamic clustering schemes. But, the repetitive formation of clusters in every round and the unclassified and random election of cluster heads initiated reasons for its static counterpart to evolve. This subsection basically covers the weight-based deterministic clustering schemes under the static category. It can be seen in following paragraphs that schemes falling in this category not only alleviate the overhead of dynamic clustering, but also the weight-based CH selection algorithms improve the overall network lifetime.

An Energy-Efficient Protocol with Static Clustering (EEPSC) for WSNs

An energy-efficient protocol with static clustering for WSN, popularly known as EEPSC was introduced by Zahmati et al. (2007) and is probably the first ever static clustering scheme. Contrary to LEACH (Heinzelman et al., 2000, 2002) in which clusters are formed in the beginning of every round, EEPSC decides the clusters only once at the beginning of network operation and continues with the same throughout the life of the network. Thus, the overhead of recurrent cluster formation is eliminated. Furthermore, in place of a probabilistic selection of cluster heads for data collection (from the cluster members), data aggregation, and transmission to base station, EEPSC engages the most eligible nodes in terms of residual energy to perform the aforementioned tasks. Since EEPSC depends on the base station for its cluster-formation, it is also categorized as a *base station assisted* clustering scheme. The network operation in EEPSC is comprised of three phases: set-up phase, responsible node selection phase, and steady state phase as follows:

Setup phase: In the setup phase, clusters are formed on the basis of distance of nodes from the base station. In order to form k clusters, the base station broadcasts k-1 messages with different powers. Nodes join the ith cluster if they hear the ith broadcast from the base station where 1<=i<=(k-1). At the end, nodes who have not responded to any of the previously broadcasted message are placed into the kth cluster. More illustratively, k=1 nodes when hearing the first message from the base station, respond to the base station through a Join-REQ message with their cluster-number set to 1. The process continues until k is set to (k-1). All remaining nodes who have not responded yet set their cluster-id equal to k through a similar Join-REQ message. Moreover, to avoid collision among the transmissions from the nodes, carrier sense multiple access (CSMA) is implemented. A new concept of temporary cluster head (TCH) has been introduced in this scheme to reduce the computational burden of cluster heads in their respective clusters. The base station selects a TCH for every cluster randomly at first and informs the respective members in every cluster. The phase ends up with a TDMA schedule from the base station defining the transmission slot for every node in the network. Clusters formed in the setup phase do not change with time and remain constant throughout the network functioning (Figure 7.7).

Responsible node selection phase: This phase starts with collecting energy status of every node (cluster-member) by the temporary cluster head (TCH). Then the TCH selects the node with maximum energy to act as the cluster head (CH) for the ongoing round and the node with the least energy as TCH for the next round. The logic behind selecting the highest energy node as cluster head is very obvious as the CH performs the tasks requiring high energy

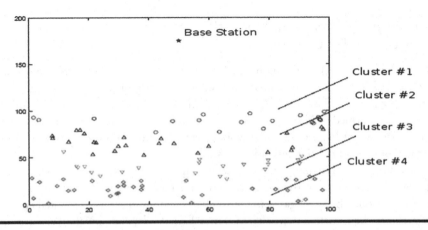

Figure 7.7   Cluster formation in EEPSC.

consumption. The newly selected (CH, TCH) is broadcasted to the cluster. The process repeats for every cluster.

Steady state phase: The steady state phase is quite similar to LEACH (Heinzelman et al., 2000, 2002). As in LEACH, cluster members transmit their sensed network information to the respective cluster heads; cluster heads in turn aggregate and forward the processed data to base station for further access by the end user.

An Enhanced Energy-Efficient Protocol with Static Clustering (E³PSC) for WSNs

EEPSC outperforms the dynamic clustering schemes viz. LEACH (Heinzelman et al., 2000, 2002), as it eradicates the overhead of cluster formation in every round along with provisioning a more eligible node as CH instead of selecting the CH in a random way. However, this scheme suffers with a major drawback that inside a cluster, if the CH is selected to be at one extreme of the cluster, the remaining nodes, especially those lying at the other extreme would have to consume more energy in communicating their data to the CH. The above mentioned problem was addressed by Chaurasiya et al. (2011) in their scheme, "An Enhanced Energy-Efficient Protocol with Static Clustering for Wireless Sensor Networks (E³PSC)" by considering the spatial distribution of deployed sensor nodes.

Being an enhancement over EEPSC, E³PSC proposes few modifications in EEPSC in order to place the CH as centrally as possible in the cluster. A brief description of the scheme, E³PSC is given below:

Setup phase: Similar to EEPSC, besides forming k-distance-based clusters in the beginning of network operation, the base station also computes the mean position of the nodes' distribution in every cluster in order to compute the relative distance of every node in the cluster from this mean position. At the end of this phase along with delivering a TDMA schedule and corresponding TCH information to the nodes in the network as in EEPSC, relative distance information from the respective clusters' mean positions is also supplied to the networks' nodes. An instance of cluster formation is given in the following Figure 7.8.

Responsible node selection phase: This phase is the heart phase of the scheme. As in EEPSC, the responsible node selection phase in E³PSC decides the CH for the current round as well as the TCH for the next round. However, not only the residual energy, but also the spatial distribution of nodes plays an important role in confirming the cluster head for the ongoing round. The node with the highest value of ($E\_Residual_i^j/D\_Mean_i^j$) becomes the cluster head of the ith cluster for the current round, and the node with the second highest value of the aforementioned expression becomes the temporary cluster head for

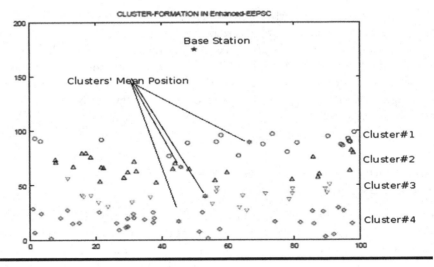

Figure 7.8   Cluster formation in E³PSC.

the next round. E_Residual$_i^j$ denotes the residual energy of the jth node ith cluster and similarly D_Mean$_i^j$ represents the distance of jth node from the mean position of nodes' distribution in ith cluster.

Steady state phase: Functionality of this phase is the same as in EEPSC. The CH collects the data from its respective cluster members and forwards the processed information to the base station for further access/use by another.

An Energy-Balanced Lifetime Enhancing Clustering (EBLEC) for WSNs

Moreover, in order to achieve further energy efficiency and improved coverage, another extension to E³PSC was proposed by Chaurasiya et al. (2012) in their scheme entitled, "An Energy-Balanced Lifetime Enhancing Clustering for WSN (EBLEC)". In this scheme, the entire network area is partitioned into a number of grids, say n =N*N/A*A, where N*N is the area of sensing field and A is the grid side which is set to be less than equal to R$_s$/2 to ensure that grid is completely under coverage even though there exists a single node in the grid as R$_s$ represents the sensing range, as shown in the Figure 7.9. In comparison to E³PSC (Chaurasiya et al., 2011) where only spatial distribution was considered to devise the most eligible node as cluster head, EBLEC considers the contribution of every node in finalizing the cluster heads for the respective clusters where the contribution of the node is computed with respect to the grid in which it lies as the ratio between node's residual energy and total residual energy of that particular grid. More illustratively, contribution is computed as the contribution factor of a node l in the grid m as follows:

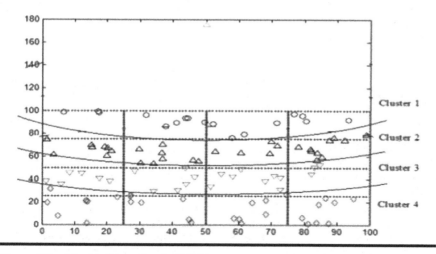

Figure 7.9   Cluster formation in EBLEC.

$$CF_m^l = \frac{E\_residual_m^l}{\sum\limits_{l=1}^{nm} E\_residual_m^l}$$

Then, the coordinate of cluster head (X_{CH}, Y_{CH}) for ith cluster is defined as the following weighted mean

$$X_{CH_i} = \left(\frac{\sum\limits_{m=1}^{N}\sum\limits_{l=1}^{nm} x_m^l * CF_m^l * I_i\left(x_m^l\right)}{\sum\limits_{m=1}^{N}\sum\limits_{l=1}^{nm} CF_m^l * I_i\left(x_m^l\right)} \right), Y_{CH_i} = \left(\frac{\sum\limits_{m=1}^{N}\sum\limits_{l=1}^{nm} y_m^l * CF_m^l * I_i\left(y_m^l\right)}{\sum\limits_{m=1}^{N}\sum\limits_{l=1}^{nm} CF_m^l * I_i\left(y_m^l\right)} \right)$$

where,

$$I_i\left(x_m^l\right) \text{ or } I_i\left(y_m^l\right) = \begin{cases} 1 & \text{if } \left(x_m^l, y_m^l\right) \in \text{Cluster}_i \\ 0 & \text{otherwise} \end{cases}$$

Once the cluster heads are decided, the scheme proceeds with the steady state phase as in its ancestor schemes—EEPSC and E³PSC.

Thus, it could be easily concluded that clustering of nodes has proved itself as a very important tool in wireless sensor networks, not only to achieve energy efficiency, but also to achieve improved coverage and connectivity, reduced delay, fault tolerance, and scalability. The importance of this tool can also be intuited because despite there being a very large number of clustering schemes in the literature, researchers are still continuing to devise new clustering-based schemes to further improve network performance.

References

I. Akyildiz, W. Su, Y. Sankarasubramaniam and E. Cayirci, A survey on sensor networks. *IEEE Communications Magazine*, vol. 40, pp. 102–114, 2002.

S. K. Chaurasiya, T. Pal and S. D. Bit, An enhanced energy-efficient protocol with static clustering for WSN. In: *Proceedings IEEE Xplore, Int'l Conf. of Information Networking (ICOIN)*, 2011, pp. 58–63.

S. K. Chaurasiya, J. Sen, S. Chatterjee and S. D. Bit, EBLEC: An energy-balanced lifetime enhancing clustering for WSN. In: *Proceedings of IEEE Xplore, 14th Int'l Conf. on Advanced Communication & Technology*, pp. 189–194, 2012.

S. Ghiasi, A. Srivastava, X. Yang and M. Sarrafzadeh, Optimal energy aware clustering in sensor networks. *Sensors Journal*, vol. 2, no. 7, pp. 258–269, 2002.

W. B. Heinzelman, A. P. Chandrakasan and H. Balakrishnan, Energy-efficient communication protocol for wireless microsensor networks. In: *Hawaii International Conference on System Sciences (HICSS)*, 2000, pp. 10–19.

W. R. Heinzelman, A. Chandrakasan and H. Balkrishnan, An application-specific protocol architecture for wireless microsensor networks. *IEEE Transactions on Wireless Communications*, vol. 1, no. 4, pp. 660–670, 2002.

N. Javaid, T. N. Qureshi, A. H. Khan, A. Iqbal, E. Akhtar and M. Ishfaq, EDDEEC: enhanced developed distributed energy-efficient clustering for heterogeneous wireless sensor networks. *Procedia Computer Science*, vol. 19, pp. 914–919, 2013.

M. Mehdi Afsar, H. Mohammad and N. Tayarani, Clustering in sensor networks: A literature survey. *Journal of Network and Computer Applications*, vol. 46, pp. 198–226, 2014.

L. Qing, Q. Zhu and M. Wang, Design of a distributed energy-efficient clustering algorithm for heterogeneous wireless sensor network. *Computer Communications*, vol. 29, pp 2230–2237, 2006.

P. Saini and A. K. Sharma, E-DEEC-enhanced distributed energy efficient clustering scheme for heterogeneous WSN. In: *1st International Conference on Parallel, Distributed and Grid Computing (PDGC)*, 2010.

F. Ye, G. Zhong, J. Cheng, S. Lu and L. Zhang, Peas: A robust energy conserving protocol for long-lived sensor networks. In: *IEEE International Conference on Distributed Computing Systems (ICDCS)*, 2003, pp. 28–37.

A. S. Zahmati, B. Abolhassani, A. A. B. Shirazi and A. S. Bahitiari, An energy-efficient protocol with static clustering for wireless sensor networks. *International Journal of Electronics, Circuit, and Systems*, vol. 1, no. 2, pp. 135–138, 2007.

TRANSPORT LAYER IV

Chapter 8

Transport Layer Caching in Wireless Sensor Networks

Introduction

A wireless sensor network consists of a number of nodes within a network which are arranged in an infrastructure-less manner that provides a route for the data for transmission through the network. The nodes are power enabled and require power for transmission of data from one sensor to another to reach to the sink node and vice versa. An increase in number of hops increases the end-to-end error rate of the network which requires an error recovery mechanism in order to keep the error rates within acceptable bounds. This end-to-end recovery of errors is the responsibility of the transport layer. The transport layer uses end-to-end retransmission to overcome the problem, and this is energy-inefficient. Hence one of the most important tasks for a wireless sensor network is to deal with reliable end-to-end transmission and optimization of power consumption within WSNs, and caching provides a well suited solution to this problem.

Transmission of data in wireless sensor network consumes more energy than data processing. Caching is a technique for quick access to data and thus can reduce overall network traffic and also optimize bandwidth utilization. Caching works on the concept that retrieving data that doesn't change directly from a sensor node is time consuming. Thus, caching stores data in sensor nodes located near the sink or at some point between the source and the sink and reduces the load on the network, and latency in retrieval of data, and it also minimizes power consumption by sensor nodes. The data is stored by using the storage capacity of sensors, and a low power antenna for communication. Since the storage capacity of sensor nodes is limited, there is scope for protocol development which can employ co-operative a caching mechanism making use of caches of nearby nodes.

With advancement in WSN, the use of WSN applications has increased. Taking the example of smart farming application of a WSN, the sensor nodes are deployed within the field for periodic measurement of soil parameters like nutrients, structure, quality, and moisture. Knowledge of these parameters make for better crop yield. To fetch information for different parameters from the field, multiple sensors are densely deployed over the field. The source node is the sensor node for collecting environmental parameters whereas the sink node collects all information from the different source nodes for processing. There is a chance that in such a dense sensor network, the position of source node to sink node can be too far and increase the probability of packet loss, and hence the requirement for end-to-end packet recovery is a network need that requires additional energy from a constrained node. This also increases the number of end-to-end retransmissions and packet latency. The solution for this is hop-by-hop recovery of packets through caching, as individual packet reliability can be maintained. Caching avoids recovery of packets all the way back to the source node, as the intermediate nodes cache the previous packet and act as new source node for sending the packet to the sink node. Thus intermediate caching increases the reliability of packets and reduces energy consumption by intermediate nodes due to end-to-end transmission.

Cache Management

The management of the cache can be: insertion of cache, elimination or replacement of cache, size of cache, location of cache, and decision on cache. The classification of cache management is shown in Figure 8.1.

Insertion Policy of Cache

It is the node that decides whether to insert a packet in cache or drop it. This classification can be done based on various factors like content mapping, duplication, neighboring node activity, traffic load, probability of packet drop, distance of communication, and cache partition.

Content mapping: The cache insertion policy depends on searching the content of intermediate nodes in cache memory. There is a header field which denotes a packet's sequence number. Each intermediate node maintains local cache progressive traffic and traffic recently transmitted. These cache entries are associated with two maps: fragment map and hole map. The fragment map contains actual cache data, whereas the hole map contains a list of missing and overdue fragments (Stann and Heidemann, 2003). There is an alternative way of processing the packet header and then forwarding the packet if the sequence number is visited for the first time. This technique requires one more header in the packet as flow ID but this increases energy consumption and fragmentation, and thus leads to poorer throughput performance.

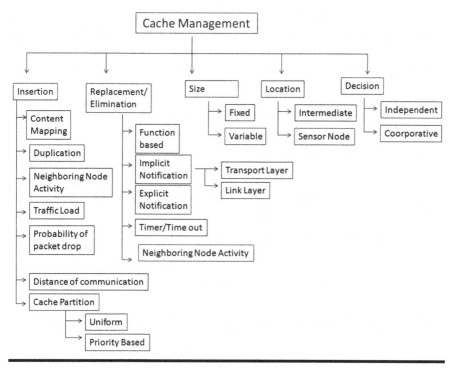

Figure 8.1   Classification of cache-based transport protocol in WSNs.

Duplication: This classification duplicates each packet received and stores it in a buffer. An intermediate node checks for each packet in its buffer, and if not found then caches the data. In this method the cache is not fully utilized, as this would fill the cache memory even if the packet cached does not need to be retransmitted. This classification can be used in applications having small amounts of traffic to send and where contention and collision are not so frequent.

Neighboring node activity: This classification uses list searching, an additional header field, and activity of valid neighboring nodes for deciding whether to packet should be cached or dropped. An intermediate node performs local retransmission by assuming that the successor node towards the destination has not yet received the cached segment. Thus, it manages a form of feedback mechanism which is useful in networks with a large number of nodes. This approach has disadvantages of requirements for additional power and computational energy for listening to neighboring nodes' activities.

Traffic load: This classification is based on factors like cache size, traffic load, and frequency of explicit acknowledgement request (EAR).

Probability of packet drop: This classification is based on the idea that old segments are given priority if not successfully send to the link layer, and priority is given to new segments if a cached packet is already acknowledged in the link layer. This is based on factor like link failure, node failure and hidden nodes.

Communication distance: Communication distance is the number of hops the packet takes in transmission, and an additional header keeps updating the packet. As the caching node is in receipt mode, the header will be checked and decide whether to cache or drop the packet. The header contains the source ID, the number of packet in a particular sequence, and distance from sink in number of hops. If the node is nearer to the sink than the sender node, then the packet is processed else dropped. An alternative approach of distributing nodes based on their distance from sink as near-source node and near-sink node is done. Thus, the probability for caching is high if the caching node is closer to the sink.

Cache partition: The cache partition reduces the work of intermediate nodes. The cache partition can be of two types: uniform and priority based. This partitioning is done because of the memory constraint of sensor nodes which will not allow all received packets from different flows to be inserted into a cache. In this categorization an intermediate node is selected as the cache point having low probability of receiving packets as flow passes through it. Thus, it allocates a bigger fraction of the cache to a longer inflow to benefit more from intermediate caching. This increases the throughput and fairness performance of the network.

Cache Replacement/Elimination Policy

Replacing or eliminating cache can be categorized into function based, implicit notification, explicit notification, timer/time out, and neighboring activity.

Function based: This approach combines different network parameters to assign a value function to each packet that is used for cache replacement. The packet with the lowest value is selected for replacement.

Implicit notification: Implicit notification uses a feedback mechanism for identifying cache packets to be eliminated. It uses ACK and NACK acknowledgement packets from the receiver for notification of lost packets. If delivery is confirmed by implicit notification, then intermediate nodes eliminate the cache entries of those packets.

Explicit notification: Cache-based transport uses explicit notification in consideration of downstream reliability in order to replace or eliminate cached packets. Explicit notification uses additional signaling packets which increase data traffic and congestion within the network.

Timer: Timer or time duration is the time at which eliminated packets are stored in the cache. Expiration of the timer can be an indicator for cache packet to decide whether they can be deleted or not. Loss detection by means of the timer by the receiver can lead to early detection of NACKs packets which may lead to an increase in traffic load and wasted energy resources.

Neighbor node activity: The activity by the neighbor node can be used to find the basis for cache replacement or elimination policy. A packet received by a successor node will be removed from the cache.

Cache Size Requirements

The cache can be divided into two parts: fixed and variable.

Fixed cache size: A watchdog timer is used for detection of loss interval and is used to determine the size of cache. Cache size is small if the watchdog timer is greater than the sending rate and vice versa. Alternatively, the size of the cache can also be a function of the probability of successful transmission of one hop or based on the probability of caching.

Variable cache size: Most of the protocol uses variable cache sizes, as most applications use heterogeneous networks where different devices have different memory allocations. Variable cache sizes adopt changing network load requirements in relation to different applications.

Cache Location

The location of cache can be of two types as intermediate and sensor node.

Intermediate cache location: Intermediate nodes which are near to sink nodes experience high packet loss due to the increase in the number of retransmission requests. Some protocols use all of their intermediate nodes to cache packets, whereas some protocols choose only specific intermediate nodes to cache packets.

Sensor node: The sensor node or source node will be used to recover data from the cache in case of the intermediate node not being able to successfully cache the data. Hence caching data at the source node provides efficiency and reliability for the limited resources of nodes.

Cache Decision

Cache decision can be independent or cooperative in nature.

Independent decision: In independent cache decision, the independent node will take the decision for caching of packets being received. Most of the protocols implement independent decision techniques by the nodes. Independent cache decision has the problem of high probability of packet redundancy, also the energy required at each node for processing and transmission is high, increasing the cost of communication.

Cooperative decision: In cooperative cache decision, multiple sensor nodes share and coordinate cache data which reduces the cost of communication and exploits aggregate cache space of the cooperating sensors. Some protocols select some sensors nodes as mediator nodes, and these represent the source node of the data to the cache nodes that store replicated data for the sink node.

Cache-Based Transport Protocols in WSNs

Protocols vary based on the following parameters: reliability direction, loss recovery, loss notification, congestion detection, congestion notification, congestion avoidance, and traffic type.

Reliability direction: The direction of reliability can be from source to sink or sink to source depending on the protocol.

Loss recovery: Loss recovery can be from end-to-end or hop-by-hop. End-to-end loss recovery involves the transport layer whereas hop-by-hop loss recovery involves the link layer. There can be a combination of both, and it depends on reliability requirements.

Loss notification: Loss notification identifies how the notification of loss takes place. The loss notification can be implicit or explicit.

Congestion detection: This parameter decides how the protocol detects congestion. There are various parameters for detection: node delay, channel probing, and buffer occupancy.

Congestion notification: Congestion notification is how the protocol notifies about congestion and can be implicit or explicit.

Congestion avoidance: This is a technique or solution for avoiding congestion within a network. It can be traffic control or resource control type. Traffic control means controlling the traffic at the source node, whereas resource control means choosing an alternative path for a congested route.

Type of traffic: This is the type of traffic that needs to be transmitted through a network. Protocols are unicast in nature, so the type of traffic supported varies by protocol.

There are many transport layer protocols proposed by different researchers for caching, some of these are discussed as follows:

Reliable multi-segment transport (RMST): This protocol was proposed by Stann and Heidemann (2003). It is a directed diffusion routing which guarantees delivery with multi-hop that exhibits a very heavy error rate. This protocol uses an end-to-end selection request NACK and hop-by-hop selective request NACK for providing end-to-end reliable services. End-to-end selection request repair requests at the sink node and travels on a reverse path from sink to source, where missing data is retransmitted, whereas hop-by-hop selective request caches fragments and forwards these fragments to the requesting node at the intermediate nodes. Thus, RMST provides two crucial services: data segmentation or reassembly and guaranteed delivery of packets using NACK-based ARQ techniques.

RMST protocol works in two modes as a caching mode and a non-caching mode. In caching mode, intermediate nodes maintain a local cache of transfers already done or planned. Whereas in non-caching mode, only the source and the sink maintain caches. In this protocol, cache entry is associated with a fragment map and a hole map. The fragment map contains actual cache data whereas the hole map consists of a list that contains missing or overdue fragments for a particular flow. As the fragment is received, the caching mode filter identifies missing or late fragments and adds them to the hole map. Data blocks are constructed at each hop, so this protocol requires high memory resources at individual nodes and is not designed to cache packets in the intermediate nodes probabilistically.

Rate-controlled reliable transport (RCRT): This protocol was proposed by Paek et al. (2007) and ensures 100% reliable delivery of segments by implementing a NACK-based end-to-end loss recovery scheme. RCRT is a reliable rate controlled transport protocol that places all congestion detection and rate adaptation functionality in the sinks. It provides better efficiency and flexibility for the network. This protocol works on the idea that the base node has more memory and can track all missing segments. The sink node detects packet loss and repairs them by requesting for end-to-end retransmission from source node. In this protocol every sensor node keeps a copy of packets sent from itself to the sink node. The sink node keep tracks of sequence numbers of packets arriving to it, any missing sequence numbers can be helpful in detecting packet loss. Thus, the sink can keep track of each packet loss within the packet; these missed packets are sent as a negative acknowledgement (NACK) from the sink node to the source node. As a NACK is received by the source node, it retransmits the requested packet to overcome the loss. This NACK is also used to avoid ACK implosion, ACKs are sent for successful reception of a segment from the sink node. Another way by which source can detect for overwrite segments of buffer by having a look at cumulative ACK sequence number piggybacked in all feedback packets.

Reliable transport memory consideration (RTMC): RTMC protocol was proposed by Zhou et al. (2008) and provides 100% reliability through hop-by-hop retransmission and congestion control. This protocol is designed in view of the limited memory of sensor nodes. As memory information in the RTMC protocol is included in packet heads and the information is exchanged among neighbors, this prevents memory overflow and overcomes the memory constraint of sensor nodes.

RTMC transmit multi-segment data from the source to the sink node through three steps: create pipe, pipe transport, and remove pipe. Create pipe creates a virtual pipe that uses an initiation packet to ask nodes to join the transport. If a node joins this transport, some memory is allocated for storing received segments. In pipe transport, these segments are transported from the virtual transport pipe. The remove pipe phase uses an end packet to tell the next hop that transport is finished. If any node buffer is empty, and it receives an end packet, then it releases memory and initializes all parameters, the virtual pipe is removed, and transport is finished for the node. These pipes are created by means of special control packets which are used at the beginning of the session and freed at the end of the session. Packets from previous node and next node are treated separately, as packets from previous node are stored in the cache and transmitted to next node as soon as possible, whereas packets from next node are used as implicit ACKs to free space in the local cache and also to determine the amount of free cache at the next node. Thus, a node will relay a packet only if there is a free cache in the next node. When a packet miss is detected, it is requested explicitly from previous node or when cache is empty. Thus, RTMC reduces transport time and makes best use of channel resources.

Actor-to-actor reliable transport protocol (A²RT): The A²RT protocol is for wireless sensor and actor networks (WSAN) where actor nodes are equipped with directional antenna and a dual radio interface. The protocol was proposed by Handigol et al. (2010). This protocol provides real time and reliable data delivery among actor nodes and sensor nodes. Since actor nodes are resource rich nodes, caching is done at actor nodes, increasing performance and efficiency of the network compared to end-to-end reliability.

A WSAN network consists of sensor nodes and actor networks, where sensor nodes are low power, low cost tiny devices which have limited sensing, computational, and wireless communicational functionality. These sensor nodes sense environmental variables and pass the information to actor nodes. Actor nodes are sensor nodes with high resources, equipped with better processing capabilities, higher transmission power, and longer/renewable battery life.

In this protocol, the first node of each partition along the route acts as a transport cache which stores packets before sending them reliably to the next transport cache along the route. A packet is removed from the cache once acknowledgement is received from the transport cache of the next actor network partition. Cross-layer interaction with the routing protocol allows a transport wrapper entity to divide the end-to-end path into multiple segments between partitioned actor nodes, minimizing the number of sensor nodes that need to be used to bridge two consecutive actor nodes. The route information is used to establish multiple transport sessions between consecutive actor partitions. Within this segment a reliable transport protocol is used for guaranteed actor-to-actor delivery of packets until the destination is reached. To assure end-to-end delivery, the end-to-end wrapper session is maintained between the source and the final destination. The source node maintains a copy of the packet which is called the master copy and which is kept until end-to-end acknowledgement from the destination node is received.

Distributed caching for sensor networks (DTSN): The DTSN protocol was proposed by Marchi et al. (2007) and is for convergecast and unicast communication in a WSN. To avoid overhead related to control and data packets, the source node keeps complete control of the loss recovery process. The loss recovery algorithm uses selective repeat ARQ using positive and negative acknowledgements for loss recovery. It uses a hop-by-hop caching mechanism to avoid end-to-end transport. This protocol is commonly used in broadband networks. It uses integration of partial buffering at source, erasure coding, and caching at intermediate nodes for achieving reliable differentiation.

DTSN is an energy efficient protocol that avoids wasting energy by optimizing control and retransmission overhead. DTSN also provides different reliability grades to fulfill the requirements of different applications in terms of latency, throughput, and energy consumption. It also supports event driven and packet driven reliability. It overcomes the limited capacities of nodes by means of exploiting the resources of the network in a cooperative manner.

DTSN implements "full reliability" by combining ACK, NACK, ARQ, and EAR (explicit acknowledgement) methods. The packets are arranged in a numbered sequence, and the transmission takes place in a session created by the source and destination node. The session is a random number created when the processing of the first packet and is terminated with a timer. The packets are stored in a buffer of source side. EAR packets are used when the source buffer is full and data is not being sent for some time. In this case an EAR packet generates a request to destination for an ACK packet. Once an ACK packet is received by the source node, it flushes its buffer. On the other hand, if a NACK packet is received, rather than an ACK packet, this shows that there are some gaps in the received data or data is not in sequence at the receiver end. In such cases, a NACK packet carries information on missing packets and also the next sequence number of the packet to be transmitted. After receiving a NACK packet, the source will retransmit the missed packet(s) followed by another EAR packet. The timer starts with sending of an EAR packet, and if the source does not get an ACK before expiration of the timer, then it generates another EAR packet. In this protocol, ACK and NACK packets are generated as a response to EAR packets and whether there are any missed packets.

Intermediate nodes upon receiving NACK packets search their cache to check for missing packets and if found, then retransmit to the destination. Intermediate nodes on receiving ACK packets clear the cache and also respond to EAR packets. The probability of caching at an intermediate node depends on size of cache, traffic load, and frequency of EAR. DTSN also implements "differentiated reliability" which is for loss-tolerant applications. In this implementation, DSTN uses a modified ARQ policy in combination with FEC strategy.

Collaborative transport control protocol (CTCP): CTCP was proposed by Li et al. (2014) and is a collaborative protocol where all nodes take part in detecting and acting on congestion control. The CTCP protocol also supports distributed storage responsibility. The CTCP protocol is scalable in nature and is independent of any underlying network layer. This protocol provides end-to-end reliability and is adaptable for different types of applications by means of variations in reliability. The protocol uses distributed fault recovery for increasing the reliability of the network and duplication of storage responsibility minimizing message loss.

In order to save energy, a two level reliability profile has been developed. First energy is saved by reducing transmissions. Second double acknowledgement is used. These reliability levels are used to provide high probability that packets reach the destination, as failure of a single node will not interrupt the data delivery. Thus, this protocol is robust even in periods of disconnect.

Pump slowly, fetch quickly (PSFQ): The PSFQ protocol was proposed by Wan et al. (2002) and is an error efficient, scalable, energy efficient, and economical protocol. In this protocol, the data transmission occurs slowly and missing segments or packets are recovered quickly from nearby nodes. The PSFQ protocol performs operation in three phases: hop-by-hop recovery, fetch/pump relationship, and multimodal operation. Hop-by-hop recovery is used for detection of losses in a hop-by-hop manner. Fetch/pump relationship is used to reduce latency by maximizing successful transfer of packets within the time frame. Multimodal operation is used to handle loss events.

The PSFQ protocol is simple and robust in nature and provides minimum signaling for reducing the cost of communication. PSFQ is customizable as per the need of the application and is highly responsive to error rates.

Hybrid and dynamic reliable transport protocol (HDRTP): The HDRTP protocol was proposed by Sharma and Aseri (2015) and is an advanced version of the PSFQ protocol, introducing additional functionality of packet signaling called inform message (INFMSG) and acknowledgment message (ACKMSG). The protocol becomes hybrid in nature by adding ACK/NACK. This protocol also overcomes the booting problem in PSFQ.

Distributed TCP caching (DTC): The DTC protocol proposed by Dunkels et al. (2004) provides direct compatibility of TCP/IP and WSN. It is implemented in the intermediate nodes and requires no change of protocol at the end nodes. It provides intermediate caching and hop-by-hop reliability. Caching is performed based on two factors: link layer feedback and the highest sequence number.

DTC protocol reduces the number of overall transmissions of TCP segments. It also decreases end-to-end retransmissions. For detection of packet loss, DTC uses timeout and duplicate acknowledgements. It uses a smaller time value for a faster response to dropped packets. ACK can be used to generate new data or loss recovery and, in order to differentiate between them, round trip time (RTT) is used. It uses SACK to set up communication among nodes and for determination of loss of packets.

The DTC protocol enhances the performance of WSNs by means of energy efficiency and throughput achieved by caching TCP segments within sensor nodes and retransmittng lost segments locally. It shifts the load from the nodes closer to the sink node or BS node, as these nodes are first to be drained of energy. Thus, this protocol uses local retransmission and segment caching and avoids end-to-end retransmission to provide better network efficiency.

Multi-path distributed TCP caching (MDTC): The MDTC protocol was proposed by Liu and Xu (2007) and increases the reliability of the transport layer by combining a multipath routing algorithm with a DTC algorithm. It uses the SPIN protocol proposed by Kulik et al. (2002) which is used to find the best route within the network by means of a master node.

In this protocol the source that wants to send packets broadcasts a routing request to the sink node through intermediate nodes. The sink node replies to this routing request using a reverse path back to the source node which is used for choosing multiple feasible paths by means of the MDTC algorithm. MDTC uses the Dijkstra algorithm for finding the best route from the source to the sink node. It uses a random selection method for finding the alternative redundant path.

It uses the sink node to make a connection with other TCP/IP networks and uses local retransmission to achieve high throughput.

Enhanced distributed TCP caching (EDTC): EDTC was proposed by Tiglao and Grilo (2012) and is an enhanced version of the DTC protocol used in CSMA/CA network with functionality of ARQ techniques. The EDTC protocol consumes less energy, provides high throughput and a better file transfer time compared to previous protocols.

EDTC uses an ARQ technique for sending a segment to the MAC buffer. It then waits for acknowledgement of the sent segment and retries for a specified number of attempts. If the maximum number of attempts is reached then the DTC module is informed of link layer transmission failure. The DTC module then locks the cache and waits for a specified time before retrying. On successful transmission and receipt of the ACK packet, the cache is unlocked, and the slot will be ready to be overwritten.

Sensor transmission control protocol (STCP): The STCP protocol was proposed by Iyer et al. (2005) and is a generic transport protocol that supports heterogeneous flow of data, congestion detection and avoidance, controlled variable reliability, scalability, and a network controlled by a base station. SCTP is an energy efficient protocol which also provides high throughput. Most of the functionality of this protocol depends on the base station (BS) which maintains a list of packets lost for which the BS has sent a NACK packet. On receiving a NACK packet, the BS clears the packet details from the record. The BS regularly and at specified intervals checks the records and if and entry is found retransmits the NACK packet.

Wisden: Wisden was proposed by Xu et al. (2004) and uses a hop-by-hop reliability scheme to improve network efficiency. All intermediate nodes in this protocol maintain a database of packets with the packet numbers. Packets are sent in a sequence with their sequence number and any gap in sequence numbers indicates packet loss. Thus, each node maintains a list of missing packets. On detection of a lost packet, a tuple with source ID and sequence number of the lost packet is inserted in the list. Entries in the missing packets list are piggybacked in outgoing transmissions, and children infer losses by overhearing this transmission. These lost packets are retransmitted in a hop-by-hop manner expect under the conditions of heavy packet loss ora change in topology, as both these cases meant there is a large number of missing packets. For updating the cache list, a simple watchdog timer is used which sets itself when the missing packet list is full and clears itself with removal of entries.

Reliable multi-segment transport (RMST): Stann and Heidemann (2003) proposed the RMST protocol which guarantees delivery for a multi hops high error rate transmission. It uses two transport layer schemes: end-to-end selective request NACK and hop-by-hop selective request NACK. In end-to-end selective request, NACK a repair request is transmitted from the sink node to a source node in a reverse path, and the source node then retransmits the data segment. With a hop-by-hop selective request NACK, a repair request is entertained by intermediate nodes which are capable of caching fragments and forward these fragments to requesting nodes.

RMST can work in two modes: caching and non-caching. In caching mode, intermediate nodes maintain local caches of traffic recently transmitted and in progress. Whereas in non-caching mode, only the sink and source maintain caches. These caches are associated with the fragment map and the hole map. The fragment map contains actual cached data that is indexed by fragment ID, whereas the hole map contains a list of missing or overdue fragments. The hole map is used by a watchdog timer which uses a timestamp to indicate when a NACK for this fragment was sent and a flag is used to determine whether a NACK is outstanding or not.

Generic information transport (GIT): The GIT protocol was proposed by Iyer et al. (2005) and uses an abstract system, information model, and reliability for supporting generalized applications. It uses a decentralized mode of functionality, and by providing modular architecture it allows different modules to reuse existing mechanisms. It reduces end-to-end delay by managing redundancy, mitigating perturbation, and reducing the number of retransmissions.

It uses a tuning and adaptation module (TAM) to recover from lost information and provide reliability. The TAM recovers from lost information through utilizing in-network spatio-temporal redundancies. Lost information is detected by the reliability module (RM), TAM then uses its reliability parameters to recover the desired information. The information is cached at each intermediate node along the path until an ACK packet is received confirming that the packet has arrived. GIT provides redundancy of information and its modular architecture can easily modify functionalities like duty cycling in order to provide an alternative route by means of sensor networks through the routing mechanism. Modular architecture enables modification in order to increase other functionalities.

End-to-end reliable and congestion aware transport layer protocol (ERCTP): This protocol was proposed by Sharif et al. (2010) and assigns priority to sensors by means of data prioritization in order to allocate bandwidth to different application and thus achieve weighted fairness. The ERCTP protocol also supports reliability and congestion control for heterogeneous WSN. It is used to detect congestion, avoid congestion, and recover lost data. It provides reliability by sensing information at a designated intermediate node for a specific period of time, and upon receiving a NACK packet resends the sensed information. Upon receipt of an ACK packet or expiration of interval, it frees up memory. It also ensures reliability by means of distributed memory storage (DMS). DMS temporary stores packet information at intermediate nodes and retransmits these packets in case the sink does not receive designated packet information. The ERCTP protocol improves throughput, reduces end-to-end packet latency for heterogeneous packet information, and reduces packet drop by means of a rate adoption mechanism.

Reliable point-to-point transport protocol (RP2PT): The RP2PT protocol was proposed by Chen et al. (2011a) and uses local transmission and caching to save energy on the network. This protocol uses broadcast messages initiated by the sink node to create a virtual circuit between the source and sink node. Nodes inside the virtual circuit cache packets with certain probabilities. The destination node uses a control window for limiting the number of received packets.

Reliable transmission protocol based on dynamic link cache (RPDC): This protocol was proposed by Chen et al. (2011b) and works by dividing intermediate nodes into near source node and near sink node based on the distance of the node from the sink and the source. This classification of intermediate nodes is dynamic in nature and can be adjusted according to the quality of links. Packet loss is checked by end-to-end signaling, and the packets are cached only once by any intermediate node.

RPDC caches packets due to two conditions. In the first condition, packets that are closer to the sink node have a higher probability of being cached in order to conserve energy in case of a NACK packet being sent when lost packets are detected by the sink node. In the second condition, packets that are closer to the source node have a higher probability of being cached as the occurrence of packet loss is lower. Thus, it can be concluded that in RPDC, for packet losses, packet caching at the source node provides more reliability than packet caching at some other intermediate nodes. Similarly, for loss of packets, caching of packets at the sink node reduces energy and time consumption.

Acknowledgement-assisted storage management (Acksis): The Acksis protocol was proposed by Wang and Wu (2007) and supports end-to-end reliability and congestion control in a hop-by-hop manner. It sends the packets in blocks, and the size of these blocks is determined based on the number of nodes within a network and the minimum storage capacity of the nodes. The Acksis protocol uses storage management which is useful for duplication checking and buffer cleaning. Duplication checking is performed by checking the received nodes against the buffer, if the packet is available in the buffer then the packet is discarded, otherwise the packet is saved. Buffer cleaning is performed by using the control message "E2ERPLY" to confirm whether the received packet is in the block. This control message is checked against the buffer, and if the packet is in the buffer, it is removed. Thus, Acksis by means of duplication checking and buffer cleaning can shorten the queue length and reduce the energy spent on transmission of duplicate packets.

Acksis performs congestion control by means of back pressure and acknowledgement packets and also contributes to storage management by means of minimizing storage overflow and increasing the delivery rate without sacrificing speed.

Dynamic acknowledgment-assisted storage management (Dacksis): Li et al. (2014) proposed the Dacksis protocol which is an advanced version of the Acksis protocol. This protocol focuses on storage management, as poor storage management may lead to node storage overflow and may degrade network performance. Dacksis focuses on end-to-end reliability, congestion control, and storage management on the network. It reduces end-to-end transmission by detecting congestion quickly. For storage management Dacksis saves node storage for unconfirmed packets which decreases the possibility of buffer overflow and packet queuing time, and also reduces the energy spent on recovery of losses. It reduces energy by sending packets in block rather than depending on per packet acknowledgement, and the size of blocks are dynamically determined.

Conclusion

This chapter discussed existing cache-based transport layer routing protocols for WSN. The protocols discussed focus on different approaches for caching in the transport layer and benefit various types of applications. It can be a solution to combine one or more surveyed protocols to advantage of their respective strengths. There are many other issues which can be discussed, and this is an open area for researchers for development of cache-based protocols for the transport layer that provide efficient and high throughput protocol.

References

H. Chen, D. Fang, X. Chen, F. Chen, X. Gong, B. Zhou and L. Qin, A reliable transmission protocol based on dynamic link cache. In: *Internet of Things (iThings/CPSCom), 2011 International Conference on and 4th International Conference on Cyber, Physical and Social Computing*, 2011a, pp. 752–755.

X. Chen, D. Fang, N. An, T. Xing, F. Chen and B. Gao, RP2PT: A reliable point-to-point transport protocol for wireless sensor networks. In: *2011 6th International Conference on Pervasive Computing and Applications (ICPCA)*, 2011b, pp. 490–496.

A. Dunkels, J. Alonso, T. Voigt and H. Ritter, Distributed TCP caching for wireless sensor networks. In: *Proceedings of the 3rd Annual Mediterranean Ad-Hoc Networks Workshop, Med-Hoc-Net*, 2004.

N. Handigol, K. Selvaradjou and C. S. R. Murthy, A reliable data transport protocol for partitioned actors in wireless sensor and actornetworks. In: *Proceedings of International Conference on High Performance Computing*, 2010, pp. 1–8.

Y. Iyer, S. Gandham and S. Venkatesan, STCP: A generic transport layer protocol for wireless sensor networks. In: *Proceedings of the 14th International Conference on Computer Communications and Networks*, 2005, pp. 449–454.

J. Kulik, W. Heinzelman and H. Balakrishnan, Negotiation-based protocols for disseminating information in wireless sensor networks. *Wireless Networks*, vol. 8, no. 2/3, pp. 169–185, 2002, ISSN 10220038.

Y. Li, R. Bartos and J. Swan, Dacksis: An efficient transport protocol with acknowledgment-assisted storage management for intermittently connected wireless sensor networks. *Pervasive and Mobile Computing*, vol. 13, pp. 272–285, 2014.

Y. Liu and K. Xu, Multi-path-based distributed TCP caching for wireless sensor networks. In: *Eighth ACIS International Conference on Software Engineering, Artificial Intelligence, Networking, and Parallel/Distributed Computing*, vol. 3, 2007, pp. 331–335.

B. Marchi, A. Grilo and M. Nunes, DTSN: distributed transport for sensor networks. In: *12th IEEE Symposium on Computers and Communications*, 2007, pp. 165–172.

J. Paek and R. Govindan. RCRT: rate-controlled reliable transport for wireless sensor networks. In: *Proceedings of the 5th international conference on Embedded networked sensor systems (SenSys '07)*. ACM, New York, NY, 305–319, 2007.

A. Sharif, V. M. Potdar and A. J. D. Rathnayaka, ERCTP: end-to end reliable and congestion aware transport layer protocol for heterogeneous WSN. *Scalable Computing: Practice and Experience*, vol. 11, no. 4, pp. 359–371, 2010.

B. Sharma and T. C. Aseri, A hybrid and dynamic reliable transport protocol for wireless sensor networks. *Computers & Electrical Engineering*, vol. 48, no. C, pp. 298–311, 2015.

F. Stann and J. Heidemann, RMST: Reliable data transport in sensor networks. In: *Proceedings of the 1st IEEE International Workshop on Sensor Network Protocols and Applications*, 2003, pp. 102–112.

N. M. C. Tiglao and A. M. Grilo, Cross-layer caching based optimization for wireless multimedia sensor networks. In: *IEEE 8th International Conference on Wireless and Mobile Computing, Networking and Communications (WiMob)*, 2012, pp. 697–704.

C. Wan, A. Campbell and L. Krishnamurthy, PSFQ: A reliable transport protocol for wireless sensor networks. In: *Proceedings of the First ACM International workshop on Wireless Sensor Networks and Applications*, 2002, pp. 1–11.

Y. Wang and H. Wu, Delay/Fault-Tolerant Mobile Sensor Network (DFT-MSN): A new paradigm for pervasive information gathering. *IEEE Transactions on Mobile Computing*, vol. 6, no. 9, pp. 1021–1034, 2007.

N. Xu, S. Rangwala, K. K. Chintalapudi, D. Ganesan, A. Broad, R. Govindan and D. Estrin, A wireless sensor network for structural monitoring. In: *Proceedings of the 2nd International Conference on Embedded Networked Sensor Systems, SenSys '04*, 2004, pp. 13–24.

H. Zhou, X. Guan and C. Wu, Reliable transport with memory consideration in wireless sensor networks. In: *2008 IEEE International Conference on Communications*, 2008, pp. 2819–2824.

Chapter 9

Congestion Control in Wireless Sensor Networks

Introduction

Wireless sensor network consists of a large number of nodes which are self-organized in nature and communicate with each other by wireless medium. The nodes in WSNs communicate through cooperation with each other by gathering data from different nodes and sending it to the central node (sink node). Data transfer from source nodes to the sink node is made in a multi-hop fashion. Factors such as resource constraint of nodes, collision of packets, node buffer overflow, contention in the transmission channel, rate of transmission, mutual interference of wireless links, many-to-one communication, and dynamic changes in network topologies are some factors that contribute to congestion in the network. Resource constraints of nodes includes factors like the limitation of a node's processing speed and limited storage capacity of sensor nodes. In many-to-one communication, the number of nodes that forward data gradually decrease with the increase in the number of forwarding hops which may result in a "funneling effect". This increase in traffic may cause congestion. Also, different event-based WSN applications like fire and disaster generate a large amount of data that can cause congestion in the network. There are many factors which contribute to congestion in the network, and this congestion needs to be avoided for fast data flow.

Congestion can occur at two level within a WSN: *node level congestion (buffer overflow)* and *link level congestion*. Node level congestion occurs when the arrival rate of packets is higher than the service rate which causes congestion due to buffer overflow of the sensor nodes. This congestion mostly occurs with nodes that are near to the sink node, as these nodes have to forward a large amount of data. This congestion directly affects network lifetime and availability. Link level congestion occurred due to collision, bit error, or competition. This congestion decreases the packet delivery rate to the sink node. Thus, to increase the efficiency, throughput, and delivery rate congestion should be avoided or handled in an appropriate manner.

There are various protocols designed by different researchers to handle congestion, and the factors that need to be kept in mind when designing congestion control protocols are throughput, average node energy consumption, scalability, goodput, fairness, packet delivery ratio, end-to-end delay, packet loss rate, and network lifetime.

Throughput: Throughput is the number of successful packets received at the sink node per time unit.

Energy consumption: The energy parameter is one of the most important parameters for WSN, as the sensor nodes always suffer from low energy due to their small size. Thus, energy an efficiency congestion control algorithm is highly desirable for development of congestion control protocols.

Scalability: The WSN needs to be scalable to support a wide range of applications. Scalability of the network may affect the performance and congestion. Thus, scalability should be kept in mind while designing congestion control protocols.

Goodput: This is a metric that indicates the best use of intermediate sensor node resources. This metric is characterized as the bandwidth delivered to all receivers.

Fairness: This is the degree of variation in the transmission rate of data.

Packet delivery ratio: This is the ratio of successfully delivered packets divided by the total number of packets.

End-to-end delay: This is the time required by the packets to reach the base station. This parameter determines the performance metric which impacts QoS in a WSN. The lower the value of this parameter the better the QoS of the network.

Packet loss rate: This is the number of packets dropped divided by the total number of packets delivered to the sink node. Packet loss rate should be low.

Network lifetime: This metric is a function of the energy consumption of the sensor nodes.

Different types of congestion control techniques can be used depending upon the nature of the WSN. The type of application for use of the WSN may impact control of the data traffic within it. Applications can classified into different types based on their types: event-driven, continuous-driven, query-driven, and hybrid.

Event-driven application: These types of WSN applications need to be designed based on the type of event that is sensed. There are many event-based applications which are mission critical and real time in nature that suddenly detect an event and generate large numbers of data packets. In these types of applications, congestion control is one of the most important parameters to ensure reliable packet delivery without delay from the source nodes to the sink node. Some examples of event-driven applications are battlefield surveillance, fire detection, target tracking, and wildlife detection.

Continuous-driven application: Continuous-driven applications are time-driven or streaming-based applications where sensor nodes continuously send data to the sink node. A continuous-driven application is used for multimedia applications like health-care monitoring and voice or video applications. These applications use constrained resources, and uncontrolled use of constrained resources increases data reporting rate, therefore, a congestion control mechanism is required.

Query-driven application: This application uses queries for data retrieval, where the sink node sends a query to the sensor nodes, and the sensor nodes respond to those queries. This operation may lead to transient congestion which needs to be addressed properly.

Hybrid applications: Hybrid applications are combination of more than one application and can be useful for advanced applications like integration of WSN with the IoT.

Congestion occurs with an increase in traffic and leads to a reduction in throughput on the network. There can be many reasons for congestion, and some are due to packet rate exceeding the service rate or conditions at link level such as contention, bit error, and interference. A congestion control mechanism can be divided into three important components: congestion detection, congestion notification, and congestion control.

Congestion detection: Congestion detection is used to detect the presence and location of the congestion. Parameters used in this protocol are buffer occupancy (queue length), channel load, and packet service time. Buffer occupancy detects congestion is triggered when the buffer occupancy of a node exceeds the constant of threshold queue length of nodes. Channel load is detects congestion by computing packet load within the network and sends an alarm when the time frame of transmission of data packet exceeds a predefined threshold. Packet service time is the time difference between arrival of packets at the MAC layer and their transmission, and it is also used to detect congestion by differentiating among these variables which is equal to one hop. There are some protocols that make use of more than one of the above parameters for congestion detection.

Congestion notification: Congestion notification means notifying the upstream nodes about when congestion is detected. There are two modes of transfer of information of congestion detection: implicit and explicit notification. Implicit notification uses piggybacking for sending notification of congestion to the upstream nodes. The congestion information is piggybacked in a payload packet header which is transmitted to upstream nodes giving information on congestion. Explicit notification uses an additional control packet to send congestion notification. Since the use of an additional control packet increases the load of an already congested network, explicit notification is rarely used.

Congestion control: Congestion is controlled by means of a protocol based on different criteria and standards for controlling congestion in the network, and these are discussed in the next section.

Congestion Control Protocols

A congestion control algorithm can be designed based on either congestion avoidance or congestion control. Congestion can be avoided by means of proper deployment of nodes, network topology, sufficient network resources, and appropriate application of network support technology. A congestion control implementation process consists of three steps: detection of congestion, notification of congestion, and handling of congestion. The congestion control algorithm can be classified into four categories depending on the congestion handling mechanism: rate regulation and allocation, routing optimization, data processing, and priority discrimination.

Congestion control based on rate regulation and allocation: Congestion can be quickly detected by means of a distributed rate regulation algorithm. Due to the complex nature of the rate regulation algorithm, it is used for small scale, mild congestion control. To overcome this limitation, centralized congestion control can achieve a good performance in global and severe congestion control.

Rate allocation is useful when the features of the network are known in advance, and rate allocation can be made at the beginning phase of the network. This technique does not work well in the case of burst data. An example of this algorithm is congestion control algorithm for a transmission path (CCF).

Rate regulation of packet flow avoids congestion. This is one of the common methods used for congestion control and can be achieved by means of regulating the forwarding rate of forwarding nodes. Speed regulation can be achieved by setting up an interval time between packets, adjusting resignation time, or adjusting the size of contention window. Examples of this protocol are ERST and IFRC.

Setting rate allocation and regulation within the network is not a good solution for real-time-based applications or to ensure QoS.

Multipath data forwarding-based congestion control: Multipath data forwarding is used to remove congestion by providing multiple paths as alternatives to the primary path in the case of congestion. There are many algorithms that can be used for determining the best paths. The path can be categorized based on algorithm implemented or by giving the priority to different paths. Some multipath protocols are congestion avoidance and congestion mechanism (MR-CACM), adaptive resource constraint (ARC) scheme, and self-selecting reliable path (SRP).

Congestion control based on traffic control: This method reduces congestion by reducing the number of packets in the network that are causing congestion. This protocol is called an additive increase multiplicative decrease (AIMD) (Chiu and Jain, 1989; Vuran and Akyildiz, 2010; Wan et al., 2009) or rate-based method. AIMD examines and checks available bandwidth and consecutively decreases the size of the congestion window to allow for a lower traffic flow. This method uses only source nodes for mitigating the congestion.

Congestion control based on routing optimization: Routing optimization uses redundancy resources for building a diversion path for alleviating congestion. This can be realized by clustering or multipath data forwarding. Routing optimization is useful for applications with a stable sending rate and a high precision requirement.

Queue assisted congestion control: This method uses the queue length of nodes as the parameter for handling congestion control. It uses a rate adjustment technique like additive increase multipartite decrease (AIMD) for keeping the length of the queue as low as possible.

Congestion control based on priority discrimination: This provides congestion control by giving priority to nodes or flow based on importance. Thus, this approach provides fairness in three aspects: the real-time requirement is different in different traffic, nodes at different levels have different priorities, the value of same data flow may be time-varying. Different protocols in this category are enhanced congestion detection and avoidance for multiple class of traffic (ECODA), congestion-aware routing (CAR), and weighted fairness guaranteed congestion control algorithm (WFCC).

Congestion control based on data processing: Congestion control is achieved by processing the data received from sensor nodes. In the case of application-oriented or event-oriented applications, data gathered at the same time from different sensor nodes or from adjacent nodes at may be redundant. This data is compressed during the transmission or sensing periods in order to alleviate congestion. Protocol support for this congestion control is greedy piecewise Constant approximation (GPCA) and data aggregation congestion control (DACC).

Clustering-based congestion control: Clustering is achieved by combining groups of sensors into a cluster with a cluster head. Clusters communicate with other clusters by means of cluster heads. Thus, the network load can be distributed by designing an efficient topology. There are many protocols proposed for cluster removal or to rotate the cluster head based on congestion. Some examples are low energy adaptive clustering hierarchy (LEACH) and sensor transmission control protocol (STCP). These protocols are discussed in Chapter 5 (Figure 9.1).

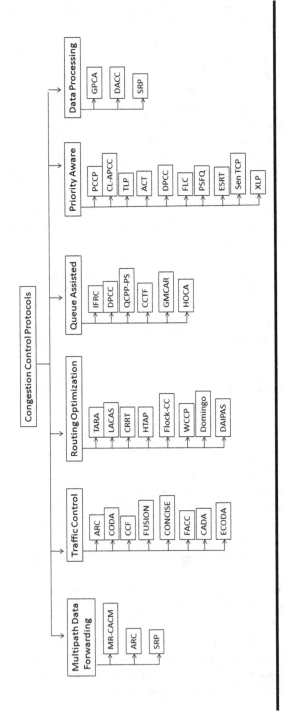

Figure 9.1 Classification of congestion control protocols.

Multipath Data Forwarding-Based Congestion Control Protocol

Multipath routing-congestion avoidance and control mechanism (MR-CACM): The MR-CACM routing protocol was proposed by Zhang (2008) and selects the best path based on the remaining energy and available bandwidth of an available path. The network state in this protocol is two-cache occupancy of nodes and the load of the channels. The protocol is used to increase the performance of reliability, decrease energy consumption, and increase the throughput of the network by assigning traffic to different paths of the network.

Adaptive resource control scheme (ARC): ARC was proposed by Kang et al. (2004) and can send traffic to alternative paths in case of congestion on the primary path, and it returns traffic to original best path as and when the path opens up. To save energy, the shunt node enters into a dormant state.

Self-Selecting reliable path (SRP): Thomas et al. proposed this protocol which provides efficient usage of bandwidth and strong tolerance to node and link failure. The next forwarding node is determined by means of radio communication. The improved version of SPR is self-healing routing (SHR) which also provides an alternative route in case the forwarder has no neighbor that is alive.

Traffic Control Protocols

Adaptive rate control protocol (ARC): The ARC protocol was proposed by Woo and Culler (2001) and improves fairness and provides an energy-efficient congestion control algorithm for WSNs. It provides a rate adjustment scheme by splitting the bandwidth between relay and locally generated packets, where preference is given to relay traffic over locally generated traffic. In this protocol, if an intermediate node overhears that packets sent previously are successfully forwarded again by its parent node, then it increases flow rate of packets. In the ARC protocol each, node estimates the number of upstream nodes and splits the bandwidth proportionally between through route and locally generated traffic. ARC avoids using control messages for congestion control.

Congestion detection and avoidance (CODA) protocol: The CODA protocol was proposed by Wan et al. (2002) and is used for detection and avoidance of congestion. It makes use of open-loop hop-by-hop backpressure and closed-loop multi-source traffic regulation for controlling the congestion on WSNs. It amalgamates the present and past loads of the channel and load level of the buffer for detecting congestion at the intermediate nodes. It makes use of an ACK packet transmitted from the sink node for regulation of the transmission of data. CODA uses the AIMD mechanism for adjusting the data flow rate. Thus, it can be concluded that CODA is a receiver-based congestion detection protocol which uses open-loop hop-by-hop backpressure and closed-loop multi-source regulations for avoiding congestion in the network. It is an energy efficient protocol which reduces the energy tax with low fidelity and eliminates flow starvation.

Congestion control and fairness (CCF) protocol: The CCF protocol was proposed by Ee and Bajcsy (2004) and uses packet service time for detecting congestion in the network. It controls congestion by controlling the rate of its downstream sensor nodes. It makes use of scalable and distributed algorithms for controlling congestion of its upstream nodes and ensuring fair delivery of packets to the BS. This protocol controls congestion in a hop-by-hop fashion where each node adjusts its rate based on the available service rate and number of child nodes. In the case of low traffic or significant packet error rate, this protocol results in low utilization. It makes use of current queue utilization which results in an increased delay in the queue, frequent buffer overflow, and increased retransmission. CCF is a distributed and scalable algorithm which does not require additional control packets and provides fair rate assignments.

FUSION: The fusion protocol was proposed by Hull et al. (2004) and detects congestion through the length of the queue to mitigate congestion. FUSION uses a hop-by-hop means of flow control mechanism, rate limitation, and prioritization MAC layer technique for handling the congestion in the network. In the instance where packets are to be dropped downstream, these packets are stopped by means of a hop-by-hop mechanism which prevents them from being transmitted due to insufficient buffer space. This is achieved by assigning priority by decreasing the random back-off timer of congested nodes. Prioritized CSMA-based MAC provides prioritization to congested nodes for accessing the channel. FUSION provides high throughput and optimized fairness. Thus, the FUSION protocol uses a rate limit and contention period reduction for handling congestion.

Congestion control from sink to sensors (CONSISE): The CONSISE protocol was proposed by Vedantham et al. (2007) and is adaptive in nature, using an adaptive rate control method for congestion control. It uses a hop-by-hop based congestion control mechanism for adjusting the downstream transmission that utilizes the maximum available bandwidth. CONSISE is a highly scalable protocol which is easy to implement and which uses its resources in an efficient manner with less overhead. It uses the concept of piggybacking, where the node receiving a packet from an upstream node will piggyback the information based on the current transmission rate and its node identifier. The downstream node upon receiving the packet updates the information as per the number of packets received from a specific node. With the expiration of a periodic timer, the node specifies its transmission rate and gives explicit feedback to the upstream node. Thus, in CONSISE a sensor node determines and adjusts the rate of sending its data based on the level of congestion at the end of each epoch.

Fairness-aware congestion control (FACC): The FACC protocol proposed by Yin et al. (2009) controls congestion and maintains fair bandwidth allocation for different flows. Nodes which are near to the source maintain a per-flow state and allocate a fair rate to each flow passing through them. Nodes which are near to the sink do not need to maintain this per-flow state, rather they use a lightweight probabilistic dropping algorithm as per the occupancy and hit frequency of the queue. The FACC protocol optimizes throughput, energy efficiency, packet loss, and fairness. Thus FACC protocol allocates a fair share of available bandwidth to each active flow according to its generated rate. The FACC protocol increases the efficiency of a network by improving the number of dropped packets, throughput, and energy consumption.

Congestion avoidance, detection and alleviation (CADA): The CADA protocol was proposed by Fang et al. (2010) and is used for detection, avoidance, and alleviation of congestion within WSNs. The CADA protocol is used for optimizing the energy consumption and information loss. It exploits a feature of data by selecting a representative sensor node from an event area as data source and then proactively mitigates and controls data traffic, and it avoids potential congestion within the network. CADA detects congestion by taking over a channel and occupation of the buffer which is alleviated by means of dynamic traffic control or source rate regulation depending upon scenario of specific hotspots. The CADA protocol optimizes throughput, energy consumption, and average end-to-end delay of the network to increase network efficiency. CADA measures the level of congestion by aggregation of buffer occupancy and utilization of the channel. CADA provides for flexible design.

Enhanced congestion detection and avoidance (ECODA): The ECODA protocol proposed by Tao and Yu (2010) is used to detect and avoid congestion. ECODA involves three mechanisms for operation. First, it uses a dual buffer threshold and weighted buffer difference for detecting congestion. Second, it prioritizes packets making use of a flexible queue scheduler. Third, for persistent congestion, it makes use of a control scheme with bottleneck node-based source transmitter rate. For transient congestion, ECODA uses hop-by-hop based congestion control for increasing network efficiency. The ECODA protocol provides energy efficient, reduced delay, better QoS, throughput, and fairness for the network.

Resource Constraint Protocol

Topology-aware resource adaptation (TARA): The TARA protocol proposed by Jaewon et al. (2007) has adopted a different strategy for multiplexing traffic which is based on specific topology. TARA adapts additional resources of the network for alleviating intersection hotspots in the case of congestion. It makes use of the parameters buffer occupancy and channel load for detecting the congestion. TARA uses distributor and merger nodes for alleviating congestion within the network where distributor nodes distribute the traffic originating from a hotspot from the original to a detour path, whereas the merger node merges two flows. TARA cannot be used for large-scale networks due to its high overhead, and it is also not suitable for scalable networks.

Learning automata based on congestion avoidance scheme (LACAS): The LACAS protocol was proposed by Misra et al. (2009) controls congestion by maintaining the balance between the arrival rate of packets and service rate of packets at each intermediate node of the network. It uses packet drop rate at the intermediate nodes as a parameter for detecting congestion within the network. LACAS makes use of previous behavior for controlling congestion efficiently. Thus, LACAS is used to avoid congestion by using a learning, automata-based approach to increase the performance of the network. LACAS is not suitable for handling link level congestion, and it has high energy consumption in the case of multi-hop paths.

Congestion-aware and rate-controlled reliable transport (CRRT): The CRRT protocol proposed by Alam and Hong (2009) increases one-hop reliability and end-to-end retransmission for loss recovery by efficiently using MAC retransmission. It centrally assigns a rate to the source by using a rate assignment policy. The CRRT protocol increases system reliability. It has high energy consumption.

Hierarchical tree alternative path (HTAP): The HTAP protocol was proposed by Sergiou et al. (2013) and uses a hop-by-hop mechanism for congestion control within the WSN. It controls congestion by making an alternative path to the sink for mitigating congestion when the primary is congested. The HTAP protocol is a simple protocol which involves minimum overhead. The HTAP protocol involves four different facets: topology control, hierarchical tree creation, alterative path creation, and handling of powerless nodes. In the topology control scheme, HTAP creates nodes and updates the neighbor table. The tree creation creates a hierarchical tree with the source node as its root. The receiver node uses a two-way handshake method to inform the transmitter node about the level of congestion on the path. In the alternative path creation, the transmitter node selects an alternative path with no congestion from the neighbor table. In handling of powerless nodes, if the battery of a sensor node is depleted, the neighbor table will be updated. This protocol employs an adaptive "congestion threshold" which enables it to avoid transient congestion situations.

Flock-based congestion control (Flock-CC): The Flock-CC protocol was proposed by Antoniou et al. (2013) and is based on the concept of swarm intelligence (behavior of birds) for designing a scalable, robust, and self-adaptive congestion control protocol for WSNs. This protocol works on concept of birds forming flocks and flowing towards a global attractor and avoiding obstacles. In the same manner, this protocol is used to guide packets to form flocks and flow towards the sink and avoid congestion within the network. The attraction and repulsion forces between packets is used to direct the motion of the flock of packets. It makes use of available network resources to dynamically balance the existing load and move packets towards the sink. This protocol also supports QoS parameters like packet loss, packet delivery ratio, energy consumption, and delay with different traffic loads.

WMSN congestion control protocol (WCCP): The WCCP protocol proposed by Mahdizadeh Aghdam et al. (2014) is a content-aware wireless multimedia sensor network that takes into account the features of multimedia content. This protocol uses the source congestion avoidance protocol (SCAP) to avoid congestion at the source side and the receiver congestion control protocol (RCCP) to avoid congestion at intermediate nodes. This protocol is used to increase network performance and video quality as received at the sink node. The SCAP protocol predicts the size of the group of picture (GoP) to detect congestion in the network and avoids congestion by adjusting the rate of transmission at the source side. The RCCP protocol detects congestion by examining and observing the length of the queue at intermediate nodes. This protocol is applicable for event-driven and monitoring traffic.

Dynamic alternative path selection protocol (DAIPaS): The DAIPaS protocol was proposed by Sergiou et al. (2014) and is a dynamic, distributed, and lightweight protocol which is used to decrease congestion in the network. In this protocol, each node of the network serves only one flow by using a soft-stage scheme to lower buffer-based congestion. If the result is not positive, through this protocol uses a hard-stage scheme for handling congestion, changing the path of data flow.

Queue-Assisted Protocols

HOCA: The HOCA protocol was proposed by Rezaee et al. (2014) and works on the idea of mitigation and congestion control using multipath routing. The HOCA protocol is a data centric congestion control protocol which uses the concept of active queue management (AQM) proposed by Athuraliya et al. (2001). This protocol was designed on parameters like end-to-end latency, network lifetime, energy consumption, and fairness. This protocol was proposed mainly for healthcare applications. This protocol avoids congestion by using multipath and QoS-aware routing. HOCA mitigates traffic by means of an optimized congestion control algorithm when congestion cannot be avoided.

The HOCA protocol prepares two routing tables: one for sensitive data and the second for non-sensitive data. Sensitive data are those data which are of high importance, whereas non-sensitive data are data about normal traffic. HOCA is divided into four phases: phase one is request dissemination performed at the sink; phase two includes the report of the even that has just occured, and this report is generated by packets that are forwarded from a sensor located at the patient's body to the sink node; phase three establishes routes; and phase four involves data forwarding and rate adjustment in the case of congestion.

Interference-aware fair rate control protocol (IFRC): The IFRC protocol was proposed by Sergiou et al. (2014), and it detects congestion by means of the length of the queue and uses a low-overhead congestion sharing mechanism for sharing the status by means of overhearing. Thus, the IFRC protocol is an interference-aware rate control mechanism. IFRC is a collection of inter-related mechanisms for distributed and adaptive fair and efficient rate allocation for tree-based communication. It makes use of AIMD control law to ensure convergence to fairness. Thus, we can say that the IFRC protocol adjusts the outgoing rate of each link based on AIMD, and this protocol provides fair bandwidth allocation to the WSN.

Queue-based congestion control protocol with priority support (QCCP-Ps): The QCCP-Ps protocol was proposed by Yaghmaee and Adjeroh (2008) and makes use of the length of queue as the parameter for determining the degree of congestion in a network. A rate is assigned to each traffic source which is based on its priority index and current degree of congestion. Thus, the QCCP-Ps protocol uses a queue-based congestion indicator which can adjust the traffic rate from source nodes based on current congestion towards upstream nodes and priority of each traffic source. QCCP-Ps is a reliable and fair transport protocol with low packet loss and is energy efficient in nature.

Grid-based multipath with congestion avoidance routing (GMCAR): The GMCAR protocol was proposed by Banimelhem and Khasawneh (2012); it is a grid- and QoS-based routing protocol and is energy efficient in nature. The GMCAR protocol uses a grid in which each grid consists of a master node which is responsible for delivering data from other nodes and routing data received from master nodes from neighboring grids. The routing table for the GMCAR protocol consists of multiple diagonal paths connecting master nodes to the sink.

This protocol checks occupancy of the buffer at the predefined path. If the buffer exceeds a predefined threshold value, then inception congestion is detected and congestion avoidance mechanisms begin which divert the incoming packet to alternative paths selected from the routing table of the master node of the grid. This is achieved by using a timer and broadcasting a message of "route invalidate."

Decentralized, predictive congestion control (DPCC): The DPCC protocol was proposed by Zawodniok and Jagannathan (2007) and includes adaptive flow and adaptive back-off interval selection schemes which work collaboratively with energy efficient, distributed power control (DPC). For detection of congestion the DPCC protocol makes use of utilization of the queue and embedded channel estimator algorithm. Once the congestion is detected, it chooses an appropriate rate by using an adaptive flow control scheme which is executed by the adaptive back-off interval selection scheme. In order to ensure fairness of weight, the weights of nodes are updated by using an optional adaptive scheduling protocol. It uses a Lyapunov-based approach to demonstrate closed-loop stability of the proposed hop-by-hop congestion control. For fair allocation of resources, the DPCC protocol uses weights that are associated with the flow of the network. In a dynamic environment, to guarantee weighted fairness, the DPCC protocol uses a dynamic weight adaptation algorithm. Thus, we can conclude that the DPCC protocol predicts congestion and is an energy efficient and decentralized protocol which works on the key concept of rate-based control and back off interval selection.

Congestion control protocol (CCTF): The CCTF protocol was proposed by Zarei et al. (2011) and makes use of a fuzzy logic system for congestion control. The CCTF protocol increases the capacity of a buffer and drops valueless packets to increase network efficiency in case of congestion.

Priority-Aware Protocols

Priority-aware congestion control protocol (PCCP): The PCCP protocol was proposed by Wang et al. (2005) and measures the degree of congestion as a ratio of packet arrival time and packet service time. The PCCP protocol achieves high link utilization and flexible fairness by means of a small buffer size, reduces packet loss, improves energy efficiency, and lowers delay. This protocol assigns priority to sensor nodes where sensor nodes with a high priority index use more bandwidth and inject more traffic. This protocol operates on single path and multipath routing. This protocol detect congestion intelligently with respect to packet inter-arrival time and packet service time. The PCCP protocol controls congestion by means of utilizing cross-layer optimization and imposing a hop-by-hop approach.

Cross-layer active predictive congestion control (CL-APCC): Cl-APCC protocol was proposed by Wan et al. (2009) and uses queuing theory to examine the flow of data from a single node as per the memory status of that node in order to increase network performance. This protocol analyze the current data change trends for forecasting and adjusting sending rate of the upcoming nodes in the next period. This protocol ensures fairness and timeliness of the WSN.

Adaptive compression-based congestion control technique (ACT): The ACT protocol was proposed by Lee and Jung (2010) and works on the key concept of reducing packet size in case of congestion by means of the adaptive compression based congestion control technique. The ACT protocol controls the queue adaptively as per the state of congestion. For compression it makes use of discrete wavelet transform (DWT), the adaptive differential pulse code modulation (ADPCM), and the run-length coding (RLC) technique.

This protocol works by changing data from time domain to frequency domain and then reduces the magnitude of data by using ADPCM. It then uses RLC to reduce the number of packets before transferring data to the source node. DWT categorizes the data based on their frequencies into four different groups, and these are also used for priority-based congestion control. ACT assigns priorities to these data groups in an inverse proportion to the respective frequencies of the data groups and defines the quantization step size of ADPCM in an inverse proportion to the priorities. RLC generates a smaller number of packets for a data group with a low priority. In the case of congestion, the ACT protocol reduces the number of packets by increasing quantization step size of ADPCM in the relaying node.

Decentralized, predictive congestion control (DPCC): The DPCC protocol was proposed by Heikalabad et al. (2011) and predicts congestion in sensor nodes and broadcasts the traffic fairly and dynamically over the entire network to prevent congestion. The DPCC protocol increases throughput, reduces loss of packets, provides fairness, and maintains low control overhead. In order to detect congestion and control it, the DPCC protocol incorporates three components as backward and forward nodes selection (BFS), predictive congestion detection (PCD) and dynamic priority-based rate adjustment (DPRA).

Fuzzy logical controller (FLC): The FLC protocol was proposed by Chen and Lai (2014) and is related to the traffic load parameter (TLP) schemes with an exponentially weighted priority-based rate control (EWPBRC). It measures the output transmission rate of the parent node and based on the amount of data transmission, it assigns an appropriate transmission rate with regards to the traffic load of each child node. This protocol is used to effectively control different rate of transmission with consumption of minimal network resources to provide efficient QoS for the network.

Pump Slowly Fetch Quickly (PSFQ): The PSFQ protocol was proposed by Wan et al. (2002) and is a simple, robust, and scalable protocol supporting different reliable data applications. This protocol works with three components: message relaying (pump operation), relay-initiated error recovery (fetch operation), and selective status reporting (report operation). It places data at a source node and distributes it in a relatively slow speed, as pump operation does. This protocol then allows fetching of missed segments from intermediate neighboring nodes in an aggressive manner, as local recovery or fetch quickly technique. However, during fetching of missed segments there is a chance that packet delivery might fail.

Event-to-sink reliable transport (ESRT): ESRT protocol was proposed by Akan and Akyildiz (2005) and is a reliable transport scheme which maintains operation in optimal operating region (OOR). This protocol provides excellent energy efficiency. ESRT protocol is self-configuring.

The ESRT protocol works by allocating the transmission rate to sensors and the sink node and centrally computes the rate of allocation. It checks for local buffer rate of the sensors, and in case of buffer overflow, sets a congestion notification bit in the packet it forwards to the sink node. The sink node receives a packet with the contention notification bit set to 1, detected congestion, and broadcasts the control signal information to all sources in order to reduce common reporting frequency. Thus, the ESRT protocol provides event-to-sink reliability to the network.

SenTCP: This protocol was proposed by Wang et al. (2005) and is an open-loop hop-by-hop congestion control protocol. This protocol uses average packet service time, average packet inter-arrival time, and buffer occupancy ratio for estimating the current degree of local congestion at intermediate nodes. The intermediate node periodically sends a feedback signal which carries the degree of local congestion and the ratio of queue length. This information is used by neighboring sensor nodes to accordingly adjust their sending rate to control congestion within the network. Thus, this hop-by-hop feedback mechanism can be used for quick removal of congestion and reduction of the packet drop rate. It is also used to conserve energy. Sen-TCP protocol increases throughput and improves energy efficiency.

Conclusion

Congestion control is one of the more important aspects in terms of providing maximum throughput and efficiency in networks, and it is one of the biggest challenges.

This chapter discussed various congestion causing factors like event-driven applications which create a heavy traffic load and may lead to congestion. Congestion may waste network energy, reduce throughput, and cause packet loss, all of which decrease network efficiency. Congestion may also increase energy consumption. This chapter discussed various congestion control techniques and protocols that use different approaches to handling congestion. Based on the study of these protocols, specific protocol can be used to overcome congestion as per specific requirements. The selection of the proper strategy and protocol for controlling congestion should be based on the applications and the location of the congestion. This chapter also discussed various factors which may lead to congestion protocols adopting different approaches to overcome with those problems. It should be noted that the issues and challenges highlighted in this chapter can be controlled and mitigated by congestion control techniques, and there is a research gap in this area.

References

O. B. Akan and I. F. Akyildiz, Event-to-sink reliable transport in wireless sensor networks. *IEEE/ACM Transactions on Networking*, vol. 13, no. 5, pp. 1003–1016, 2005.

M. M. Alam and C. S. Hong, CRRT: Congestion-aware and rate-controlled reliable transport in wireless sensor networks. *IEICE Transactions on Communications*, vol. 92, no. 1, pp. 184–199, 2009.

P. Antoniou, A. Pitsillides, T. Blackwell, A. Engelbrecht and L. Michael, Congestion control in wireless sensor networks based on bird flocking behavior. *Computer Networks*, vol. 57, no. 5, pp. 1167–1191, 2013.

S. Athuraliya, S. H. Low, V. H. Li and Q. Yin, REM: Active queue management. *IEEE Network*, vol. 15, no. 3, pp. 48–53, 2001.

Y.-L. Chen and H.-P. Lai, A fuzzy logical controller for traffic load parameter with priority-based rate in wireless multimedia sensor networks. *Applied Soft Computing*, vol. 14, pp. 594–602, 2014.

D.-M. Chiu and R. Jain, Analysis of the increase and decrease algorithms for congestion avoidance in computer networks. *Computer Networks and ISDN Systems*, vol. 17, no. 1, pp. 1–14, 1989.

C. T. Ee and R. Bajcsy, Congestion control and fairness for many-to-one routing in sensor networks. In: *Proceedings of the 2nd International Conference on Embedded Networked Sensor Systems*, 2004.

W.-w. Fang, J.-m. Chen, L. Shu, T.-s. Chu and D.-p. Qian, Congestion avoidance, detection and alleviation in wireless sensor networks. *Journal of Zhejiang University Science C*, vol. 11, no. 1, pp. 63–73, 2010.

S. R. Heikalabad, A. Ghaffari, M. A. Hadian and H. Rasouli, DPCC: Dynamic predictive congestion control in wireless sensor networks. *International Journal of Computer Science Issues*, vol. 8, no. 1, pp. 472–477, 2011.

B. Hull, K. Jamieson and H. Balakrishnan, Mitigating congestion in wireless sensor networks. In: *Proceedings of the 2nd International Conference on Embedded Networked Sensor Systems*, 2004, pp. 134–147.

K. Jaewon, Z. Yanyong and B. Nath, TARA: Topology-aware resource adaptation to alleviate congestion in sensor networks. *IEEE Transactions on Parallel and Distributed Systems*, vol. 18, no. 7, pp. 919–931, 2007.

J. Kang, Y. Zhang, B. Nath and S. Yu, Adaptive resource control scheme to alleviate congestion in sensor networks [C]. In: *Proceedings of Workshop on Broadband Advanced Sensor Networks (BASENETS)*, 2004.

J.-H. Lee and I.-B. Jung, Adaptive-compression based congestion control technique for wireless sensor networks. *Sensors*, vol. 10, no. 4, pp. 2919–2945, 2010.

S. Mahdizadeh Aghdam, M. Khansari, H. R. Rabiee and M. Salehi, WCCP: A congestion control protocol for wireless multimedia communication in sensor networks. *Ad Hoc Networks*, vol. 13, no. Part B, pp. 516–534, 2014.

S. Misra, V. Tiwari, M. Obaidat. (LACAS): learning automata based congestion avoidance scheme for healthcare wireless sensor networks. *IEEE Journal on Selected Areas in Communications*, vol. 27, no. 4, 466–479, 2009.

A. A. Rezaee, M. H. Yaghmaee, A. M. Rahmani and A. H. Mohajerzadeh, HOCA: Healthcare Aware Optimized Congestion Avoidance and control protocol for wireless sensor networks. *Journal of Network and Computer Applications*, vol. 37, pp. 216–228, 2014.

C. Sergiou, P. Antoniou and V. Vassiliou, A comprehensive survey of congestion control protocols in wireless sensor networks. *IEEE Communications Surveys & Tutorials*, vol. 16, no. 4, pp. 1839–1859, 2014.

C. Sergiou, V. Vassiliou, and A. Paphitis. Hierarchical Tree Alternative Path (HTAP) algorithm for congestion control in wireless sensor networks. *Ad Hoc Networks*. 11(1), 257–272, 2013.

L. Q. Tao and F. Q. Yu, ECODA: Enhanced congestion detection and avoidance for multiple class of traffic in sensor networks. *IEEE Transactions on Consumer Electronics*, vol. 56, no. 3, pp. 1387–1394, 2010.

R. Vedantham, R. Sivakumar and S.-J. Park, Sink-to-sensors congestion control. *Ad Hoc Networks*, vol. 5, no. 4, pp. 462–485, 2007.

M. C. Vuran and I. F.Akyildiz, XLP: A cross-layer protocol for efficient communication in wireless sensor networks. *IEEE Transactions on Mobile Computing*, vol. 9, no. 11, pp. 1578–1591, 2010.

C.-Y. Wan, A. T. Campbell and L. Krishnamurthy, PSFQ: A reliable transport protocol for wireless sensor networks. In: *Proceedings of the 1st ACM International Workshop on Wireless Sensor Networks and Applications*, 2002, pp. 1–11.

J. Wan, X. Xu, R. Feng and Y. Wu, Cross-layer active predictive congestion control protocol for wireless sensor networks. *Sensors*, vol. 9, no. 10, pp. 8278–8310, 2009.

C. Wang, K. Sohraby and B. Li, SenTCP: A hop-by-hop congestion control protocol for wireless sensor networks. In: *Proceedings of the IEEE INFOCOM*, 2005, pp. 107–114.

A. Woo and D. E. Culler. A transmission control scheme for media access in sensor networks. In: Proceedings of the 7th annual international conference on Mobile computing and networking (MobiCom '01), pp.221–235, 2001.

M. H. Yaghmaee and D. Adjeroh, A new priority based congestion control protocol for wireless multimedia sensor networks. In: *Proceedings of the 2008 International Symposium on World of Wireless, Mobile and Multimedia Networks (WoWMoM)*, 2008, pp. 1–8.

X. Yin, X. Zhou, R. Huang, Y. Fang and S. Li, A fairness-aware congestion control scheme in wireless sensor networks. *IEEE Transactions on Vehicular Technology*, vol. 58, no. 9, pp. 5225–5234, 2009.

M. Zarei, A. M. Rahmani and R. Farazkish, CCTF: Congestion control protocol based on trustworthiness of nodes in wireless sensor networks using fuzzy logic. *International Journal of Ad Hoc and Ubiquitous Computing*, vol. 8, no. 1, pp. 54–63, 2011.

M. Zawodniok and S. Jagannathan, Predictive congestion control protocol for wireless sensor networks. *IEEE Transactions on Wireless Communications*, vol. 6, no. 11, pp. 3955–3963, 2007.

J. Zhang, Congestion avoidance and control mechanism for multi-paths routing in WSN. *International Conference on Computer Science and Software Engineering*, vol. 5, pp. 1318–1322, 2008.

HETEROGENEOUS WIRELESS SENSOR NETWORKS

V

Chapter 10

Architecture, Advances, and Challenges

Like any other ad hoc network, WSNs can also be defined as peer-to-peer networks which consist of a large number of sensor nodes each with the functionality viz. data collection, storage, processing, and forwarding. With time, WSN has emerged as a very useful tool for applications such as environment sensing, habitat monitoring, disaster prediction and management, weapons control, and intelligent transportation that not only ease human life but also bless social living with a great bliss. Being an application specific network, WSNs may use independent network structures to support different applications. To improve the quality of service (QoS), a WSN may involve a number of specialized nodes which result in a different variant of network called a heterogeneous wireless sensor networks (HWSN). Such wireless networks consists of sensor nodes which are equipped with different abilities in functionality, sensing and computational abilities, and battery power. For example, there may be an application scenario where the WSN may contain nodes of different sensing ranges, different sensing tasks, or of different computational abilities. We can also intuit that the mixed deployment of such sensor nodes may result in a network which is much better balanced in performance and cost, and for this is reason the HWSN is emerging as a very popular variant of wireless sensor networks. There are basically two kinds of nodes in the heterogeneous wireless sensor networks: regular nodes and super nodes. Both classes of nodes have the equal access to the network but have different functions to perform. Where the regular nodes are meant for environment monitoring/sensing and forwarding the collected trust information to the super nodes, the super nodes are capable of performing specialized processing and connecting to other networks via gateway nodes (Wang, 2013). The node heterogeneity in HWSN has already been exploited in many research proposals to achieve a number of goals viz. network lifetime enhancements.

On the basis of node heterogeneity, heterogeneous wireless sensor nodes can be categorized in the following three major categories:

1. Energy heterogeneity based HWSN
2. Computational heterogeneity based HWSN
3. Link heterogeneity based HWSN

Energy heterogeneity based HWSN: In such heterogeneous wireless sensor networks, nodes may be equipped with differently powered batterie. It may also refer to the networks where sensor nodes have replaceable batteries.

Computational heterogeneity based HWSN: This class of HWSN refers to the networks where nodes are equipped with different processing capabilities, i.e., some of the nodes have more powerful processors along with more storage capacity with respect to other nodes in the network, thus enabling such nodes to have added computational power.

Link heterogeneity based HWSN: In such networks, nodes are heterogeneous in the sense that some nodes have long-distance and highly-reliable communication links with respect to others. Link heterogeneity is provided by varying transmission power levels of different nodes.

Being a variant of the distributed networks, HWSNs are autonomous and multihop networks allow the nodes to join and leave the networks at any instance of time resulting in a dynamic network. This characteristic of such networks necessitates provisioning self-organization to ensure network connectivity at all times.

In some applications, different types of catastrophic data may be required to be collected from t emergency situations or from some harsh environment which necessitates the development of a data collection protocol for such networks where node are heterogeneous in nature. Also, in large-scale deployment, reducing nodes' energy consumption concludes in a prolonged network lifetime and, the same has drawn the attention of many researchers to device energy-efficient protocols almost at every layer to keep the network alive for a longer period. Along with these, several issues which pertain to normal sensor networks, i.e., wireless sensor networks with homogeneous nodes, viz. minimizing end-to-end delay and provisioning medium access for the nodes are also need to be dealt with.

This chapter presents a detailed typical architecture of the heterogeneous wireless sensor networks (HWSN) and surveys the major cluster-based routing schemes in its subsequent sections, as clustering has already proved its significance in dealing with the aforementioned issues in the context of homogeneous wireless sensor networks.

Network Architecture

A heterogeneous wireless sensor network is envisioned to consist of a large number of nodes which are of different abilities in sensing, computing, communicating, and power, and this may result in various types of traffic such as scalar and multimedia data when compared to traditional wireless sensor networks. As in traditional WSNs, nodes are homogeneous in nature, i.e., they all have the same physical capabilities generating only a single kind of simple scalar data which has enabled the development of scalable network architecture for such networks.

However, the inherent heterogeneity caused by the technologies casting the different types of sensor nodes and the support for various kinds of data traffic, require a completely different approach with respect to the same in traditional WSNs. More illustratively, the large volume of traffic coming to nodes with different capabilities may not be supported by topologies consisting of nodes with limited processing capabilities. Similarly, due to the high processing/computation nature, nodes in heterogeneous WSNs may require high power batteries with respect to the need of traditional ones due to their low processing abilities. Also, the simultaneous processing and data transfer may not be supported by each device in the homogeneous network.

The intrinsic heterogeneity of the nodes in the network leads to a more realistic class of networks—single-tier HWSNs and multi-tier HWSNs. Wherein a single-tier heterogeneous wireless sensor network, lies a flat topology of nodes with different abilities, in the multi-tier HWSNs, capabilities of nodes viz. processing capabilities, sensing capabilities, and storage capabilities are exploited to form a hierarchical network operation.

Single-Tier Architecture

As briefed above, in the single-tier architecture, multiple nodes of the network form a flat topology where the nodes are equipped with different processing powers, sensing and transceiving powers, and storage abilities. In such architecture, there may be a peer working among the nodes or it may be hierarchical in nature. In the peer HWSN, every node in the network works in an independent manner and can send data directly to the sink for further processing as shown in Figure 10.1(a), whereas, in the hierarchical single-tier architecture, a subset of the nodes deployed may have further responsibility for performing local processing with respect to the other nodes deployed in the same network as described in Figure 10.1(b). Such nodes are referred to as cluster heads serving a particular group of nodes. The information gathered and processed by such local processing hubs or cluster heads is then transmitted to the sink for further processing and use.

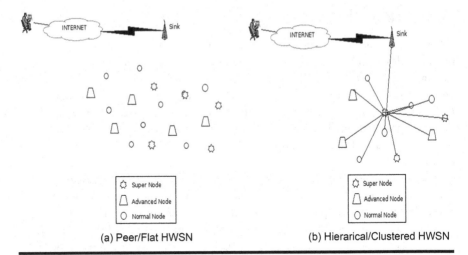

(a) Peer/Flat HWSN

(b) Hierarical/Clustered HWSN

Figure 10.1   Single-tier network architecture for HWSN: (a) Peer/Flat HWSN; (b) Hierarchical/Clustered HWSN.

Multi-Tier Architecture

In multi-tier architecture, based on nodes' abilities, the network is layered to facilitate adaptive use of network resources. In this network architecture, nodes are arranged to form hierarchies based on their abilities, i.e., the lower end scalar nodes lie at the bottom of the hierarchy to perform simpler tasks, and the higher end specialized nodes are placed at the higher up to perform more complex tasks as shown in Figure 10.2. Moreover, each hierarchy in this multi-tier architecture may

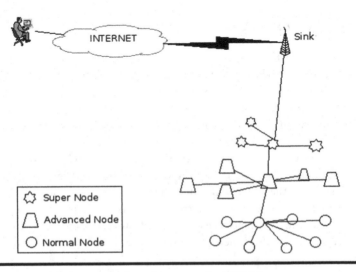

Figure 10.2   Multi-tier network architecture for HWSN.

implement clusters also with some node acting as cluster heads. Such provisioning of clusters and cluster heads may reduce the data traffic being communicated to the higher layers, as the cluster heads at their respective tiers may aggregate the data being transported.

Advances in HWSN

With respect to the two very popular applications, tracking and monitoring, of wireless sensor networks in the context of heterogeneity, most of the work is subjected to the efficient data transmission from the field to base station, i.e., how the data can be routed to the base station in the most efficient way.

HWSN is comprised of nodes having diverse hardware along with different initial energy and thus, it provides support to network lifetime enhancement without any significant increase in the cost (Al-Karaki and Kamal, 2004). However, it must be taken care that the nodes rich in resources must not be overburdened, and network load must be divided into appropriate proportions among the nodes in the network.

Since, the nodes are resource-constrained, and energy is the most precious resource, energy heterogeneity is researched the most for HWSNs. Moreover, as in homogeneous wireless sensor networks, clustering has proved itself a very significant tool in achieving scalability, fault-tolerance, energy-efficiency, improved coverage, and connectivity. A number of clustering based routing schemes have already been proposed in the heterogeneous variant of sensor network and will be discussed.

In the available literature, two types of heterogeneous sensor network are investigated the most: 2-level and 3-level and above.

In the 2-level energy heterogeneous model, two types of nodes—normal and advanced—are deployed in the field with different initial energy. Similarly, in a 3-level energy heterogeneous model, three types of nodes—normal, advanced, and super—are deployed with different initial energy. Thus, the available clustering based routing schemes can be categorized belonging to one of two categories—2-level or 3-level.

2-level energy heterogeneity is adequately dealt with in DEEC, distributed energy-efficient clustering, algorithm (Qing et al., 2006) and DDEEC, developed distributed energy-efficient clustering (Elbhiri et al., 2010). 3-level energy heterogeneity is dealt with accordingly in EDEEC, enhanced distributed energy efficient clustering scheme (Saini and Sharma, 2010); DDEEC, enhanced developed distributed energy-efficient clustering; and BEENISH, balanced energy efficient network integrated super (Qureshi et al., 2013).

All the schemes mentioned above can be treated improvements to LEACH (Heinzelman et al., 2000, 2002), as the cluster head selection process is similar to that in LEACH. Traditionally in LEACH, network operation is broken into rounds.

Each round is comprised of two phases—cluster formation and steady state phase. In the cluster formation phase, cluster heads are selected on the basis of a threshold say $T(s)$, computed as per the suggested percentage of cluster heads, known a priori, say p, and whether a node has been elected as cluster head in the previous $(1/p)$ round or not as per the following formula (Heinzelman et al., 2000, 2002):

$$T(s) = \begin{cases} \dfrac{p}{1 - p \cdot \left(r \bmod \dfrac{1}{p} \right)} & \text{if } s \in G \\ \\ 0 & \text{Otherwise} \end{cases}$$

Here, r is the current round and G is the set of nodes that have not been elected as cluster head for the last $(1/p)$ rounds.

Nodes generate a random number between 0 and 1, and if the number is less than $T(s)$, they declare themselves as cluster heads and the process goes on.

Since in heterogeneous wireless sensor networks, nodes differ in their initial energy, care must be taken to incorporate the changed scenario and for the same modification to be made in defining the reference value of p for different types of nodes in the schemes—DEEC, DDEEC, EDEEC, EDDEEC, and BEENISH respectively.

2-Level Energy Heterogeneity

In a heterogeneous wireless sensor network with 2-level energy heterogeneity, two types of nodes, say normal and advanced nodes, with different initial energy are deployed in the network where initial energy of the advanced nodes is increased by a factor, a, with respect to that of normal nodes, i.e., if E_0 is the initial energy of normal nodes and $(E_0 + aE_0)$ is of advanced nodes. Moreover, if N is the total number of nodes in the network, a fraction, say m, of this N is deployed as advanced nodes, i.e., a total mN advanced nodes are deployed along with N(1-m) normal nodes. Hereby, the total initial energy of the network is $N * E_0 * (1 + m * a)$.

Distributed energy-efficient clustering algorithm for heterogeneous wireless sensor networks (DEEC) was envisioned by Qing et al. (2006) and was developed as a modification of LEACH (Heinzelman et al., 2000, 2002) in order to handle the needs of heterogeneous sensor networks.

In DEEC, the probability used to elect the cluster heads is defined as the ratio between the residual energy of nodes and average residual energy of the network as follows (Qing et al., 2006):

$$p_i = \begin{cases} \dfrac{p_{\text{opt}} E_i(r)}{(1 + am)\bar{E}(r)} & \text{if } s_i \text{ is the normal node} \\ \\ \dfrac{p_{\text{opt}}(1 + a)E_i(r)}{(1 + am)\bar{E}(r)} & \text{if } s_j \text{ is the advanced node} \end{cases}$$

E_i is the residual energy of ith node and E' is the average residual energy of the network defined as follows:

$$\overline{E}(r) = \frac{1}{N} \sum_{i=1}^{N} E_i(r)$$

Similar to LEACH, cluster heads are elected as per the above cited expression and nodes get associated with an appropriate cluster head based on the received signal strength, and clusters are formed. Later on, in the steady transmission phase, as in LEACH (Heinzelman et al., 2000, 2002), cluster members send their sensed data to the respective cluster head, which in turn sends the aggregated data to the sink for further processing at the user end.

Moreover, the solution by Qing et al. (2006) was proposed to extended to multi-level energy heterogeneity.

Developed distributed energy-efficient clustering for heterogeneous wireless sensor networks (DDEEC): DDEEC was developed by Elbhiri et al. (2010). The concept behind DEEC is to engage the most eligible nodes, those with higher residual energy, as a cluster head. The scheme penalizes advanced nodes by selecting them as cluster heads again and again because the probability for advanced nodes to be chosen is higher with respect to that of normal nodes. However, it is very possible that the advanced nodes come closer to normal nodes in terms of their residual energy after being some time on the network. Thus, DEEC becomes a non-optimal solution.

Elbhiri et al. (2010) proposed a modified probability expression to handle the situation with an idea of a threshold energy (residual) below which both normal as well as advanced nodes would be treated equally as follows:

$$p_i = \begin{cases} \dfrac{p_{opt} E_i(r)}{(1 + am)\overline{E}(r)} & \text{Nml nodes, } E_i(r) > Th_{REV} \\[4mm] \dfrac{(1 + a)p_{opt} E_i(r)}{(1 + am)\overline{E}(r)} & \text{for Adv nodes, } E_i(r) > Th_{REV} \\[4mm] c\,\dfrac{(1 + a)p_{opt} E_i(r)}{(1 + am)\overline{E}(r)} & \text{for Adv, Nml nodes, } E_i(r) \leq Th_{REV} \end{cases}$$

Th_{REV} is the threshold residual energy value and is defined as bE_0, b being a constant E_0 being the initial energy of the node. It can be easily intuited that if set to 0, the scheme behaves like DEEC (Qing et al., 2006). Moreover, it has been found through a number of experiments that when b is set to 0.7, the scheme gives the best result. Therefore,

$$Th_{REV} = 0.7 * E_0$$

Similarly, c is another constant which directly controls the number of clusters. From the above mentioned expression, it can be easily grasped that the larger the value of c, the more clusters would be formed. Through an extensive set of experiments, the most suitable value of c is found to be 0.2 by Elbhiri et al. (2010).

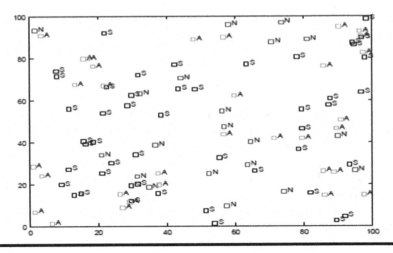

Figure 10.3 Node deployment in 3-level HWSN.

3-Level Energy Heterogeneity

As an extension to the 2-level heterogeneous wireless sensor network, energy heterogeneity was increased to incorporate an additional category of nodes called the super nodes. These are equipped with an extra $b*E_0$ initial energy when compared to the normal nodes and advanced nodes in 2-level heterogeneous networks, where b is a constant between 0 and 1. Also, N is the total number of nodes and m the fraction of N which are non-normal nodes, i.e., advanced and super nodes. If m_0 is the fraction of mN which are super nodes, then a total mm_0N super nodes are there in the network and $mN(1-m_0)$, $N(1-m)$ are the number of advanced and normal nodes respectively. Therefore, the total initial energy of the network is $N*E_0*(1+m*(a-m_0*a+m_0*b))$ (Figure 10.3).

Enhanced distributed energy efficient clustering protocol for heterogeneous wireless sensor networks (EDEEC): This protocol was devised by Saini and Sharma (2010) as an extension to DEEC (Qing et al., 2006) and DDEEC (Elbhiri et al., 2010) to include another category of nodes called super nodes as explained above. Similar to DEEC and DDEEC, the probability computing expression was adjusted accordingly as follows (Saini and Sharma, 2010):

$$p_i = \begin{cases} \dfrac{p_{opt}E_i(r)}{(1+m.(a+mo.b))\overline{E}(r)} & \text{if } s_i \text{ is the normal node} \\[2ex] \dfrac{p_{opt}(1+a)E_i(r)}{(1+m.(a+mo\cdot b))\overline{E}(r)} & \text{if } s_j \text{ is the advanced node} \\[2ex] \dfrac{p_{opt}(1+b)E_i(r)}{(1+m.(a+mo.b))\overline{E}(r)} & \text{if } s_i \text{ is the super node} \end{cases}$$

And accordingly, the threshold computation for normal, advanced, and super nodes can be made as:

$$
T\left(s_i\right)=\begin{cases}
\dfrac{p_i}{1-p_i \cdot \left(r \bmod \dfrac{1}{p_i}\right)} & \text{if } p_i \in G' \\[4ex]
\dfrac{p_i}{1-p_i \cdot \left(r \bmod \dfrac{1}{p_i}\right)} & \text{if } p_i \in G'' \\[4ex]
\dfrac{p_i}{1-p_i \cdot \left(r \bmod \dfrac{1}{p_i}\right)} & \text{if } p_i \in G''' \\[4ex]
0 & \text{Otherwise}
\end{cases}
$$

where G', G'', and G''' are the set of normal, advanced, and super nodes respectively which have not been elected as cluster heads for the respective values of $(1/p_i)$. Immediately after the cluster heads are finalized, clusters' formation takes place and the network's intended task of tracking/monitoring starts.

Enhanced developed distributed energy efficient clustering protocol for heterogeneous wireless sensor networks (EDDEEC): EDDEEC was proposed by Javaid et al. (2013) to deal with 3-level heterogeneous sensor networks and is basically an improvement over the EDEEC (Saini and Sharma, 2010) in order to tackle the issues introduced in DDEEC (Elbhiri et al., 2010). Elbhiri et al. in their scheme for DDEEC drew attention to the fact that after some network operation rounds, it might be very possible that the residual energy of extra capable nodes comes closer to that of normal nodes, then overburdening such nodes with respect to normal nodes may serve as an injustice to them and create inappropriate load distribution among the nodes. The probability formula for cluster head selection in EDDEEC is given as follows:

$$
p_i=\begin{cases}
\dfrac{p_{opt}E_i(r)}{\left(1+m\left(a+m_o b\right)\right)\bar{E}(r)} & \text{for } N_{ml} \text{ nodes} \\
& \text{if } E_i(r) > T_{\text{absolute}} \\[2ex]
\dfrac{p_{opt}(1+a)E_i(r)}{\left(1+m\left(a+m_o b\right)\right)\bar{E}(r)} & \text{for Adv nodes} \\
& \text{if } E_i(r) > T_{\text{absolute}} \\[2ex]
\dfrac{p_{opt}(1+b)E_j(r)}{\left(1+m\left(a+m_o b\right)\right)\bar{E}(r)} & \text{for Sup nodes} \\
& \text{if } E_i(r) > T_{\text{absolute}} \\[2ex]
c\dfrac{p_{opt}(1+b)E_i(r)}{\left(1+m\left(a+m_o b\right)\right)\bar{E}(r)} & \text{for } N_{ml}, \text{Adv}, \text{Sup nodes} \\
& \text{if } E_i(r) \leq T_{\text{absolute}}
\end{cases}
$$

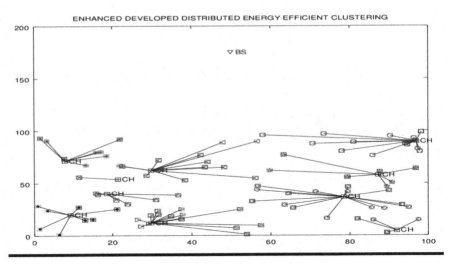

Figure 10.4   An instance of cluster formation in EDDEEC.

$T_{absolute}$ is a new constant provisioned in the scheme in order to achieve a balanced load distribution when residual energy of advanced nodes and super nodes comes close to that of normal nodes. $T_{absolute}$ is defined as zE_0 where z is the experimental constant set to 0.7 and E_0 is the initial energy of normal nodes.

Like its predecessor schemes, EDDEEC proceeds with cluster formation and data transmission phases to achieve the intended network functionality (Figure 10.4).

Balanced energy efficient network integrated super heterogeneous protocol (BEENISH): Qureshi et al. (2013) proposed the scheme BEENISH to include another level of heterogeneity by introducing ultra super nodes which are equipped with u*E_0 with respect to normal, advanced, and super nodes. Then, accordingly, the probability formula is enhanced as follows:

$$p_i = \begin{cases} \dfrac{p_{opt}E_i(r)}{\left(1+m\left(a+m_0\left(-a+b+m_1(-b+u)\right)\right)\right)\bar{E}(r)} & s_i \text{ is the normal node} \\[2em] \dfrac{p_{opt}(1+a)E_i(r)}{\left(1+m\left(a+m_0\left(a+b+m_1(-b+u)\right)\right)\right)\bar{E}(r)} & s_i \text{ is the advanced node} \\[2em] \dfrac{p_{opt}(1+b)E_i(r)}{\left(1+m\left(a+m_0\left(-a+b+m_1(-b+u)\right)\right)\right)\bar{E}(r)} & s_i \text{ is the super node} \\[2em] \dfrac{p_{opt}(1+u)E_i(r)}{\left(1+m\left(a+m_0\left(-a+b+m_1(-b+u)\right)\right)\right)\bar{E}(r)} & s_i \text{ is the ultra-super node} \end{cases}$$

Based on this probability computation, a round specific threshold is computed accordingly as follows:

$$T(s_i) = \begin{cases} \dfrac{p_i}{1 - p_i \left(r \bmod \dfrac{1}{P_i} \right)} & \text{if } s_i \in G \\ 0 & \text{otherwise} \end{cases}$$

Nodes generating random numbers less than $T(s_i)$ declare themselves as cluster heads and the remaining nodes join them to form clusters. After cluster formation, members send their sensed data to the respective cluster head, which in turn forwards the processed/aggregated data to the base station destined for the end user.

Through an extensive set of experiments, BEENISH was proven to be a supreme way to handle level 4 and above energy heterogeneity.

Future Challenges in HWSN

With ongoing technological advances, heterogeneity is becoming an intrinsic feature of the networks deploying sensors for various applications. Many researchers have been working towards improving the experience of such heterogeneous networks involving sensors since the very advent of this domain. A HWSN consists of a large number of sensor nodes of different capabilities and improves the communication experience in such networks. This is a very vast challenge, especially with regard to the self-organization and data collection, data aggregation, hardware design, and scalability etc.

Design of Smart Self-Organizing and Data Collecting Protocols

Self-organization refers to the ability of nodes to get engaged with one another in order to establish a network despite their random deployment. Since power and computational resources are limited in sensor networks, achieving efficiency in self-organization of the network is becoming a very challenging task for researchers.

Moreover, with the evolution of the internet of things (IoT), deployment of heterogeneous nodes has increased a lot, and since the rate of change in topology is very high in sensor networks, care must be taken while providing solutions for data collection so that the nodes can adjust dynamically as per the changing topology.

Design of Efficient Data Aggregation Scheme for HWSN

The *data aggregation* process refers to fusion of data packets from different nodes in order to produce a single packet data. The data fusion strategy is relatively simple when the network contains homogeneous nodes only as the data being collected at the processing end is similar in nature. However, it is not the case with heterogeneous sensor networks, as the nodes with different abilities may generate data of dissimilar nature and fusing them becomes quite a tough job for the researchers. Hence, design of efficient data aggregation schemes to meet the requirement of HWSN is a major challenge.

Design of Hardware for HWSN

Nodes in traditional wireless sensor networks are resource-constrained in nature, i.e., they come up with limited computational, processing, and communication abilities. Also, they are equipped with limited power which cannot be easily replaced when there are harsh and inaccessible environments.

But to meet the current industrial and business application requirements, such limited capacities are insufficient. Hence, designing hardware while keeping the intended application requirements in mind is a new challenge before the research community.

References

J. N. Al-Karaki and A. E. Kamal, Routing techniques in wireless sensor networks: A survey. *IEEE Wireless Communications*, vol. 11, no. 6, pp. 6–28, 2004.

B. Elbhiri, R. Saadane, S. El Fkihi and D. Aboutajdine, Developed distributed energy-efficient clustering (DDEEC) for heterogeneous wireless sensor networks. In: *5th International Symposium on I/V Communications and Mobile Network (ISVC)*, 2010.

W. B. Heinzelman, A. P. Chandrakasan and H. Balakrishnan, Energy-efficient communication protocol for wireless microsensor networks. In: *Hawaii International Conference on System Sciences (HICSS)*, 2000, pp.10–19.

W. R. Heinzelman, A. Chandrakasan and H. Balakrishnan, An application-specific protocol architecture for wireless microsensor networks. *IEEE Transactions on Wireless Communications*, vol. 1, no. 4, pp. 660–670, 2002.

N. Javaid, T. N. Qureshi, A. H. Khan, A. Iqbal, E. Akhtar and M. Ishfaq, EDDEEC: enhanced developed distributed energy-efficient clustering for heterogeneous wireless sensor networks. *Procedia Computer Science*, vol. 19, pp. 914–919, 2013.

L. Qing, Q. Zhu and M. Wang, Design of a distributed energy-efficient clustering algorithm for heterogeneous wireless sensor network. *Computer Communications*, vol. 29, pp. 2230–2237, 2006.

T. N. Qureshi, A. H. Khan, A. Iqbal, E. Akhtar and M. Ishfaq, BEENISH: Balanced energy efficient network integrated super heterogenous protocol for wireless sensor networks. *Procedia Computer Science*, vol. 19, pp. 920–925, 2013.

P. Saini and A. K. Sharma, E-DEEC-enhanced distributed energy efficient clustering scheme for heterogeneous WSN. In: *1st International Conference on Parallel, Distributed and Grid Computing (PDGC)*, 2010.

X. B. Wang, Heterogeneous architecture for ad hoc networks. In: *Advanced Materials Research*, vol. 756, Trans Tech Publ., 2013, pp. 1059–1062.

Chapter 11

QoS Provisioning at the MAC Layer in Heterogeneous Networks

Introduction

With the introduction of low power sensors and microcontrollers, the use of WSN technology has manifold uses. A WSN consists of many sensors in the network that communicate hop-by-hop to the sink node or base station. The sink node works as a processing unit which process these data for the required result. The deployment of these nodes can be in a random fashion or predetermined. Energy conservation is one of the prime considerations, as nodes are battery operated and battery drainage is a major issue. This chapter focuses on increasing the lifetime of a network by maintaining energy conservation within the entire network in which heterogeneous nodes are being used.

A heterogeneous wireless sensor network (HWSN) consists of sensor nodes of different functions for the type of sensing needed and communication range, providing greater flexibility in deployment. Heterogeneity means that heterogeneous nodes have more powerful microprocessors and memory than the normal nodes. Thus, the powerful computational capacity of heterogeneous nodes can be useful in processing complex data and for long-term storage. There are basically two kinds of nodes in heterogeneous wireless sensor networks: regular super nodes. Both classes of nodes have equal access to the network but have different functions to perform. Where the regular nodes are meant for environment monitoring/sensing and

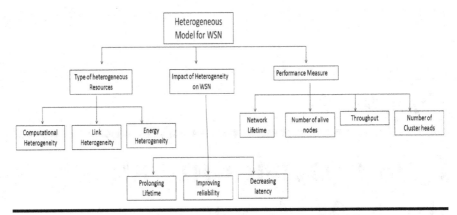

Figure 11.1   Heterogeneous model for WSNs.

forwarding the collected trust information to the super nodes, the super nodes are capable of performing specialized processing and connecting to other networks via gateway nodes (Wang, 2013). The node heterogeneity in HWSN has already been exploited in many research proposals to achieve a number of goals viz. network lifetime enhancements.

Even the clustering technique can be of two types: homogeneous and heterogenous depending on whether applied to a WSN or a HWSN. The energy saving schemes for homogenous and heterogeneous network perform differently on the two types of network.

The heterogeneous model for WSN can be defined as in Figure 11.1.

The resource heterogeneity of a heterogeneous network can be classified into three groups as:

1. Computational
2. Link
3. Energy

Computational heterogeneity based: This class of HWSN refers to networks where nodes are equipped with different processing capabilities, i.e. some of the nodes have more powerful processors along with more storage capacity with respect to other nodes and so have added computational power.

Link heterogeneity based: In such networks, nodes are heterogeneous in the sense that some nodes have long-distance and more reliable communication links with respect to other nodes. Link heterogeneity is provided by varying transmission power levels of different nodes.

Energy heterogeneity based: In such heterogeneous wireless sensor networks, nodes may be equipped with batteries of different powering. It may also refer to the networks where sensor nodes have replaceable batteries.

Energy heterogeneity supersedes the computational and link heterogeneity, as computational heterogeneity and link heterogeneity consume more energy.

Heterogeneity within WSN can be impacted by the following factors:

1. Prolonging network lifetime
2. Improving reliability of data communication
3. Decreasing latency of data transportation

Prolonging network lifetime: The average energy consumption for forwarding a packet from a sensor node to the sink node is less in heterogeneous network than in homogenous network.

Improved reliability of data communication: Links between sensor nodes have lower reliability and each hop within the network lower end-to-end delivery rate. Heterogeneous networks work on decreased number of hops between sensor nodes and the sink node and thus provide a higher end-to-end delivery rate than homogenous sensor network.

Decreasing latency of data transportation: Computational heterogeneity can decrease the processing latency in intermediate nodes and link heterogeneity can decrease the waiting time in the transmitting queue.

Performance measures are used for calculating performance of protocols in WSN as follows:

1. Network lifetime
2. Number of cluster heads per round
3. Number of alive nodes per round
4. Throughput

Network lifetime: Network lifetime is the time duration from the start of the operation of sensor network to the death of the first alive node.

Number of cluster heads per round: This is the number of nodes which send the aggregated information from their cluster member directly to the sink node or base station.

Number of alive nodes per round: This is the number of nodes that have not yet expended all of their energy.

Throughput: This is the total amount of data sent over the network, which includes rate of data sent from a cluster head to the base station and from nodes to the cluster head.

MAC Layer

WSNs use a five-layered OSI architecture.

Application Layer
Transport Layer
Network Layer
Data Link Layer
Physical Layer

These five layers are controlled by three cross layers for efficient working of WSNs, these layers are power management, connection/mobility management, task management layers.

Application layer: The application layer is used for the management of traffic and provides software for various types of applications. These applications are used to send queries for obtaining the information from the system.

Transport layer: This layer provides internet work communication. This layer uses multiple protocols for providing reliability to the system and avoiding congestion within the network. As wireless sensor network is based on multihop transmission of data, we generally do not use a TCP-based connection. UDP connection is more desirable in WSNs.

Network layer: The network layer supports various types of routing protocols. Routing protocols are used to maintain various aspects like power consumption, memory, reliability, and redundancy types of factors. Thus, working on routing protocols is one of the prominent areas of work for researchers in WSNs. Routing protocol can be based on factors like flat routing, hierarchical routing, event driven, query driven, or time driven based on the type of application in which the WSN is being deployed. To provide redundancy of data, WSNs use data aggregation or data fusion. Data aggregation combines data from different nodes and changes it into meaningful information to reduce the need for redundancy of data and thus saves energy. Data fusion is a more advanced concept of data aggregation which is used to remove noise from aggregated data.

Data link layer: This layer provides reliability of data from point-to-point to point-to-multipoint. This layer also supports error control and multiplexing of data. This layer supports MAC addresses which are hardware addresses of nodes. This layer is used to provide higher reliability, low delay, higher efficiency, and throughput.

Physical layer: This layer provides the interface for transmission of the data stream over the physical medium. This layer deals with the frequency of data transmission that includes selection of frequency, generation of carrier frequency for modulation, signal detection, and security.

MAC is defined as sub-layer of data link layer of WSN shown above. The main responsibilities of the MAC layer are channel access policies, scheduling, buffer management, error control, and arbitration of access to one channel shared by all nodes. MAC layer protocols for WSNs need to be energy efficient to increase the lifetime of network. MAC protocols are used to consider factors like energy efficiency, reliability, high throughput, and low access delay in order to accommodate limited resources of sensors and lower the power consumption by sensor nodes.

Quality of Service (Qos)

Quality of service in terms of customer perspective can be defined as the ability of a service provider to maintain a specific level of service agreement, termed as service level agreement (SLA). This SLA is used for the classification of traffic from customers, and once the traffic is being classified, it is marked for differentiated services using different protocols. In QoS implementation, the traffic is classified and can partition the network traffic into different levels of priority and class of services.
QoS can be classified into two different types:

1. Differentiated services (DiffServ)
2. Integrated services (IntServ)

Differentiated services: The DiffServ model is RFC-2475, which allocates resources by dividing the traffic into a small number of classes.

Intergrated services: This can be used for real-time applications like remote videos, multimedia conference, etc. This QoS is RFC-1633, which provides end-to-end signaling. It uses "resource reservation" and "admission control" for implementing QoS. RSVP is used for signaling explicitly the need for QoS for traffic in end-to-end paths within the network (Dumka et al., 2018).

QoS within WSNs can be classified into two groups as:

1. Application-specific
2. Network-specific

Application-specific: In application-specific, QoS only focuses on quality of application. In this type of QoS, quality of application depends on the number of sensor nodes, quality of sensing, deployment of nodes, lifetime, coverage, resolution of cameras, etc.

Network-specific: In network-specific, QoS focuses on quality of service at the time of delivery of data by the communication protocol stack. This can achieve QoS by means of efficient use of resources at each layer for providing optimized energy efficiency, latency, throughput, packet loss, etc.

Contribution of MAC Layer Toward QoS

There are various methods by which the MAC layer can contribute to QoS directly or indirectly, and this section of the chapter discuss properties of these mechanism and how they contribute to QoS.

Service Differentiation: One of the best techniques or methods used for providing QoS is service differentiation (Dumka et al., 2018). By means of service differentiation within a wired or wireless network, the user requirement can be achieved. This method prioritizes and differentiates different traffic by means of classification of traffic. These different classes are treated in different manners based on their degree of importance and on the requirement for resource allocation within different classes. Service differentiation consists of two phases as follows:

1. Priority assignment
2. Differentiation between different priority levels

Priority assignment is used to assign priority among different traffic of the network based on their importance within the network. Priorities can be assigned in three different ways:

1. Static priority assignment
2. Dynamic priority assignment
3. Hybrid priority assignment

In the case of static priority assignment, priority is assigned once the packet is created and the priority does not change until the packet reaches its destination. There can be different parameters for static priority assignment:

Traffic class: Traffic class prioritizes the traffic based on the class of traffic, as in this case traffic can be of different types like real-time traffic, non-real time or best effort traffic, and so these traffic patterns are classified into different classes based on the demands of traffic within the network (Tomar et al., 2015).

Static priority: This assignment can be made by the type of source which generates the sensed data from sources. In this type of assignment, a particular source node or sink node can be assigned a priority based on the requirement of that node or a priority can be assigned based on the distance of that node from the sink. Packets generated from the node will inherit the propertiesd of its creator. Data can be prioritized based on the event-based model discussed earlier in this chapter.

In *dynamic priority assignment*, the priority of the node is assigned on a real-time basis, and the priorities can be changed based on the requirements of nodes and conditions within the network. The different criteria for assigning priorities within dynamic priority assignments are as follows:

Remaining hop count: This is based on older age being given priority. In this type of priority assignment, the packets which are far away from the destination in terms of hop count, where hops refer to number of nodes, are given more priority than packets which are nearer to the sink or destination node. This is based on the fact that packets which are further from the sink or destination node are more prone to be dropped due to deadline miss or timer duration miss.

Traversed hop count: This factor is used to prioritize the traffic based on the facts that packets which traverse a higher hop count should be given priority over the packets which traverse a lower hop count in order to reach the sink node. Packets which traverse higher hop counts are more prone to drop or miss deadline. Thus, giving priority to such packets enchances the QoS by optimizing network resources, channel utilization, and increasing network lifetime and delivery ratio.

Traffic load: Traffic load can also be used to prioritize traffic. There are different types of nodes within a WSN: leaf node, relay node, and cluster node. Giving priority to the sensor node that is has a heavier forwarding load can reduce the rate of packet drop within the network. This is because heavier forwarding loads will overflow the buffer of the receiving node and cause packet loss. Thus prioritizing such node traffic will enhance the QoS by reducing packet loss.

Hybrid priority assignment: This method uses a combination of static and dynamic priority assignment as per the requirement of the network service delivery. This is much better than both, as this assignment is mostly needed with the current requirement of WSN to provide efficient QoS.

Differentiation method: Differentiation method is used for successful sharing of resources within the WSN as per the requirements of service delivery. This stage is the succeeding step after assigning priority in the network. Resource sharing can be done using different traffic classes at the MAC layer which can be categorized as follows:

Changing contention window size: This technique is used for setting priority within the network where there are multiple sending nodes which are sharing a single communication channel for sending their data concurrently. In order to avoid any collision among different nodes, the contention window is adopted where different contender nodes request channel allocation, and one of the nodes wins contention and reserves the communication channel for sending its data. This node reserves a contention period for sending the data and after finishing the data transmission or the end of the contention period, the next node gets the chance to reserve the communication channel for sending its data. Now to extend this medium sharing among different traffic classes with high priority as per the needs and requirements of service delivery, the contention window size can be changed or reduced to prioritize and give chances to the high priority traffic classes. If a node with a high priority traffic class gets the contention window for the communication channel, then the contention window size of such node will be increased to give a longer time period to send high priority traffic. Thus, by changing contention window size, efficient QoS can be achieved. Contention window selection based on non-uniform probability distribution can also contribute to achieving efficient QoS as normal selection is on a random basis.

Changing inter-frame space (IFS) duration: Inter-frame space duration is the space or the time for which the sensor node stays quiet before contention or back-off timer. Changing the IFS duration based on different types of traffic class according to priorities assigned to them can enhance the QoS of the network.

Scheduling of transmission slot: Scheduling can be done based on prioritizing traffic and demand of traffic for efficient delivery.

There are different protocols used in MAC layer for *error control* which can accommodate service differentiation by changing the persistency of transmission or strength of error control codes. Error resiliency of traffic belongs to a different priority of classes which can be controlled easily and thus can be used to provide QoS for traffic.

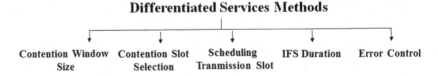

Differentiated Services Methods

| Contention Window Size | Contention Slot Selection | Scheduling Tranmission Slot | IFS Duration | Error Control |

Additional factors include:

Error control: Error control is needed for providing efficient and reliable delivery of packets within the network and can be used to optimize energy dissipation. Error control can be achieved through three mechanisms:

1. ARQ technique
2. Forward Error Correction (FEC)
3. Hybrid ARQ technique

ARQ technique can use a stop and wait or sliding window protocol for error control which can control errors by means of acknowledgement packets. The acknowledgment tells the sender of successful transmission of a packet. The use of either protocol will depend on the type of usage.

Forwarding error control (FEC) uses different techniques for controlling the errors within the network. This technique uses redundant bits which are added to the sending data to make sending data more secure. These redundant bits are checked at the receiving side, and after confirmation are removed to fetch the actual data at the receiving node.

Hybrid ARQ uses ARQ and FEC as an error control mechanism. The sender uses a simple FEC code to encode the packet, and if the receiver gets an error message, then it sends a negative acknowledgement, and the sender uses a stronger FEC code to convert the message into an encoded form. This process continues until the receiver gets a correct packet.

Different Qos-Based Protocols for Heterogeneous Networks

In order to provide QoS in a heterogeneous network, there are many protocols proposed by researchers with added features to make these protocols more efficient as follows.

There is a protocol proposed, the AMPH protocol, which uses the concept of the Z-MAC protocol in a heterogeneous network. This protocol hybridizes the functionality of contention-based and schedule-based approaches for maximizing utilization of channels in heterogeneous WSNs. AMPH uses time division multiplexing where a node can transmit at any slot to increase utilization of a channel and minimize latency on the network. This protocol assigns priority to traffic with high priority over other traffic in the time slot. The AMPH protocol supports real-time and best-effort traffic, where real-time traffic is given preference over best-effort traffic. This traffic class is statically set up in the application layer.

This protocol operates under two phases: setup and transmission. Setup phase initiates an action like neighbor discovery, slot assignment, framing, and synchronization. It uses the DRAND algorithm for slot assignment. Once the setup phase is done, nodes go to transmission phase. Transmission phase deals with processing transmission, where the node sends packets until either its queue is empty or the time slot has expired.

The AMPH protocol is designed for heterogeneous WSN which provides service differentiation, high throughput, and support for QoS. This protocol provides real-time traffic latency and prevents best-effort traffic starvation. It also provides high reliability by using time division scheduling thus supporting efficient delivery of heterogeneous traffic for a new generation of applications providing efficient QoS.

Dilipand and Patel (2009) proposed an energy efficient heterogeneous clustered (EEHC) protocol for heterogeneous networks which is used for selection of a cluster head in a distributed fashion within a heterogeneous WSN. The selection of a cluster head is based on residual energy of the node with respect to other nodes within the network. The node with higher residual energy is selected for the cluster head. This algorithm is based on the LEACH algorithm used for homogenous WSNs.

Duan and Fan (2007) proposed a distributed energy balancing clustering (DEBC) protocol for heterogeneous WSN which applies a cluster-based approach, where a cluster head is selected based on the ratio of remaining energy of a node to average energy of the network. The node having a larger value of the ratio is selected as the cluster head for processing of data of the cluster. This protocol also considers 2-level heterogeneity, and then it extends the results for multi-level heterogeneity.

Qing et al. (2006) also proposed a protocol called distributed energy efficient clustering scheme (DEEC) for a cluster-based approach. This protocol also selects the cluster head based on the ratio of remaining energy of a node to average energy of the network The epochs acting as cluster heads for nodes are determined according to their initial and residual energy. In order to have uniform energy, the DEEC protocol rotates the cluster head among all of the nodes. Two levels of heterogeneous nodes are considered in the algorithm, and after that a general solution for multi-level heterogeneity is obtained. In order to gain global knowledge of a network, DEEC estimates the ideal value of network lifetime which is used to determine the reference energy each node expends during a round. The DEEC protocol provides a better result in terms of network lifetime and effectiveness.

Elbhiri et al. (2010) proposed a developed distributed energy efficient clustering (DDEEC) scheme for heterogeneous wireless sensor networks which dynamically and effectively changes probability of election of cluster head. This protocol is an advanced version of DEEC protocol, which uses which uses initial and residual energy of all the nodes to find the cluster head. This protocol also like DEEC protocol estimate ideal value of network lifetime which is used for determining the reference energy each node expend during round. In this protocol, cluster head collects information from nodes of the cluster and transmit aggregated data directly to base station.

Rashed et al. (2011) proposes weighted election protocol (WEP) which is energy efficient protocol and enhances the stability period of WSN. This protocol combine the clustering strategy with chain routing algorithm which satisfy energy and stable period constrains under heterogeneous environment in wireless sensor networks. This protocol assign weights to each node which is determined by dividing the initial energy of each node by initial energy of normal node. It then uses approach like LEACH for selecting the cluster head within the cluster. In order to connect different cluster head, this protocol uses greedy algorithm which forms chain among selected cluster heads. Once the chain has been formed, a chain head is selected randomly among all cluster head nodes. The cluster head nodes in each cluster then fused those data and finally send to the base station.

Elbhiri et al. (2009) proposed an improved version of DEEC called stochastic distributed energy efficient clustering (SDEEC) for heterogeneous WSNs which is a self-organized network with dynamic clustering concept. This protocol dynamically selects a cluster head and the criteria for selection of a cluster head is residual energy of nodes. This protocol allocates a transmission time to each node, and each node has to send data to the cluster head within its transmission time frame. The cluster head receives all data from different nodes and compresses the data into a single signal by means of signal processing. After compression, the cluster head sends the consolidated data to the prime cluster head. The non-cluster heads save energy by changing their states to sleep mode.

Saini and Sharma (2010) proposed the threshold distributed energy efficient clustering (TDEEC) protocol which uses an energy efficient cluster head scheme for HWSN. This protocol uses a threshold value for deciding on the cluster head which is based on residual energy and average energy of that round in respect to the optimum number of cluster heads.

Marin et al. (2008) proposed an energy efficient service discovery protocol (C4SD) for heterogeneous wireless sensor networks which offers distributed storage of service descriptions. This protocol assigns a unique hardware identifier and weight to each node which is called a capability grade, and the higher degree of a capability grade the more probability it has for becoming the cluster head. These nodes act as a distributed directory of service registrations for the nodes in the cluster. The structure ensures low construction and maintenance overhead and reacts rapidly to topological changes of the sensor network by making decisions based only on the 1-hop neighborhood information, thus avoiding chain-reaction problems. A service lookup results in visiting only the directory nodes and ensures a low discovery cost.

Kumar et al. (2010) proposed the distributed cluster head election (DCHE) for HWSN. It selects the cluster head based on weighted probability. The cluster's member nodes communicate with the elected cluster head, and then the cluster heads communicate the aggregated information to the base station via single-hop communication. In this protocol, the authors classified nodes based on battery energy capacity as type-1, type-2, and type-3 nodes. These nodes are distributed uniformly over the network.

Kour and Sharma (2010) proposed a heterogeneous-hybrid energy efficient distributed protocol (H-HEED) for HWSN. This protocol assumes that the node population is equipped with more energy than the rest of the nodes in the same network which creates heterogeneity in terms of node energy. It selects the cluster head based on residual energy of each node. This protocol divides the nodes based on their energy into 2-level, 3-level, and multilevel.

Jing et al. (2005) proposed the cluster-based energy balancing scheme for HWSN which classifies nodes as "strong" in terms of abundant storage, computing and communication abilities as well as energy. A strong node is selected as the cluster head which collects information from other sensor nodes of the cluster by means of multi hop links and communicates with the base station through single hop links. For communication among cluster heads, cluster heads form a connected backbone. The communication between regular nodes and the cluster head is done with low transmission power, whereas communication between cluster heads is done in the high transmission range. Thus, these two types of traffic are carried on different frequency bands or encoding techniques and making for better utilization of energy resources.

Du and Lin (2005) proposed the cluster head relay (CHR) routing protocol for heterogeneous sensor networks which forms a network by using two types of sensors with a single sink. These two types of sensors are a large number of low-end sensors, denoted by L-sensors, and a small number of powerful high-end sensors, denoted by H-sensors. These sensors determine their location by using some location service like GPS and are static in nature. These two sensors are uniformly distributed over the network. The protocol partitions a heterogeneous network into clusters, each being composed of L-sensors and led by an H-sensor. Within a cluster, L-sensors collect data of environmental variables and forward the data to the cluster head in a multi hop fashion. On the other hand, H-sensors fuse data within their own cluster and forward aggregated data packets originating from other cluster heads toward the sink in a multi hop fashion using only cluster heads. The communication from L-Sensor to H-Sensor within the same cluster is done using short range data communication, whereas communication between H-sensor to sink is done using long range data communication.

Li (2010) proposed an energy efficient cluster head election protocol for heterogeneous wireless sensor networks using the improved Prim's algorithm to construct an inter cluster routing. This protocol proposes a heterogeneous protocol using three types of sensors based on their energy resources. The cluster head in this protocol sets up a TDMA schedule and transmits this schedule to the other nodes in the cluster. Thus, this protocol avoids collision among data messages. The non-cluster head node saves energy by turning itself off except during transmitting time. This protocol uses a multi hop routing protocol for cluster heads which are far away from the base station in order to save energy.

Xuegong et al. (2009) proposed the cluster multi-hop transmission (CMHT) protocol for HWSN. This protocol uses weight value for selection of the cluster head within a cluster and transfers data by using nodes in the cluster and the cluster head in multi hop transmission. This protocol balances energy consumption by avoiding the death of the cluster head due to rapid energy drainage and thus prevents the situation of cluster chain block that is caused by failure of one cluster head. This protocol extends the stable phase of data communication time by extending the time of each cycle and thus reducing the number of cyclical cluster re-establishment. This protocol also reduces frequency of cluster head election, as frequent cluster head election costs energy and so prolongs survival time of the network. This protocol prevents excessive energy consumption by using cluster and cluster head multi-hop transmission for long-distances and thus better uses energy on the entire network.

Conclusion

This chapter presented the HWSN and QoS parameters associated with it. The heterogeneous network gives a better performance in terms of lifetime and network reliability. Techniques like clustering further enhance the performance of the network. This chapter discussed various factors which affect the performance of heterogeneous networks and also different protocols discussed by different scholars in the field.

Related Work

Akyildiz et al. (2002) discusses that standard MAC protocols proposed for wireless networks cannot be used for sensor networks, and different sensor based MAC protocols with different objectives has been proposed with the main objective of maximizing network lifetime. Demirkol I. et al. (2006) presented a survey for commonly used MAC protocols for WSN. They proposed different protocols for different types of applications. Different protocols serve different requirements as per the applications. Teng and Kim (2010) and Ullah et al. (2010) presented a survey of different application specific MAC protocols like delay sensitive, mission critical, and bandwidth hungry. Application specific characteristics like interactivity and reliability affect network design.

Saxena et al. (2008) discusses different QoS aware routing protocols for WSN. These protocols focus on accommodation of the different types of QoS-constraint traffic and adaptation for varying traffic loads.

References

J. Ai, D. Turgut and L. Bölöni, A cluster-based energy balancing scheme in heterogeneous wireless sensor networks. In: *Proceedings of the 4th International Conference on Networking (ICN'05)*, 2005, pp. 467–474.

I. F. Akyildiz, W. Su and Y. Sankarasubramaniam and E. Cayirci, Wireless sensor networks: A survey. *Computer Networks*, vol. 38, pp. 393–422, 2002.

I. Demirkol, C. Ersoy and F. Alagöz, MAC protocols for wireless sensor networks: A survey. *IEEE Communications Magazine*, vol. 44, no. 4, pp. 115–121, 2006.

X. Du and F. Lin, Improving routing in sensor networks with heterogeneous sensor nodes, *Proceedings of IEEE 61st Vehicular Technology Conference*, vol. 4, 2005, pp. 2528–2532.

C. Duan and H. Fan, A distributed energy balance clustering protocol for heterogeneous wireless sensor networks. In: *Wireless Communications, Networking and Mobile Computing, International Conference on*, 2007, pp. 2469–2473.

A. Dumka, "Innovation in software defined network and network function virtualization" IGI Global, 2018.

A. Dumka. IoT based traffic management tool with hadoop based management scheme for efficient traffic management. *International Journal of Knowledge Engineering and Data Mining*, 2018.

A. Dumka, H. L. Mandoria, V. Fore and K. Dumka, Implementation of QoS Algorithm in the Integrated Services (IntServ) MPLS Network. In: *Proceedings of 2nd International Conference on Computing for Sustainable Global Development (INDIAcom)*, pp. 1048–1050, 2015.

B. Elbhiri, R. Saadane and D. Aboutajdine, Stochastic distributed energy-efficient cluster-ing (SDEEC) for heterogeneous wireless sensor networks. *ICGST-CNIR Journal*, vol. 9, no. 2, p. 1117, 2009.

B. Elbhiri, R. Saadane, S. El fldhi and D. Aboutajdine. Developed distributed energy-efficient clustering (DDEEC) for heterogeneous wireless sensor networks. In: *5th International Symposium On I/V Communications and Mobile Network*, pp. 1–4, 2010.

Md. Golam Rashed, M. Hasnat Kabir and S. E. Ullah, WEP: An energy efficient proto-col for cluster based heterogeneous wireless sensor network. *International Journal of Distributed and Parallel Systems*, vol. 2, no. 2, pp. 54–60, 2011.

A. Jing, D. Turgut, and L. Bölöni. A Cluster-based energy balancing scheme in heterogeneous wireless sensor networks. *Lecture Notes in Computer Science*. 3420. 467–474, 2005.

H. Kour and A. K. Sharma, Hybrid energy efficient distributed protocol for heterogeneous wireless sensor network. *International Journal of Computer Applications*, vol. 4, no. 6, pp. 1–5, 2010.

D. Kumar, T. C. Aseri and R. B. Patel, EEHC: Energy efficient heterogeneous clus-tered scheme for wireless sensor networks. *Computer Communications*, vol. 32, pp. 662–667, 2009.

D. Kumar, T. C. Aseri and R. B. Patel, Distributed cluster head election (DCHE) scheme for improving lifetime of heterogeneous sensor networks. *Tamkang Journal of Science and Engineering*, vol. 13, no. 3, pp. 337–348, 2010.

H. Li, An energy efficient routing algorithm for heterogeneous wireless sensor networks. *Proceedings of the International Conference on Computer Application and System Modeling (ICCASM 2010)*, pp. V3-612–V3-616.

R. S. Marin-Perianu, J. Scholten, P. J. M. Havinga and P. H. Hartel, Cluster-based ser-vice discovery for heterogeneous wireless sensor networks. *International Journal of Parallel, Emergent and Distributed Systems*, vol. 23, pp. 1–35, 2008.

L. Qing, Q. Zhu and M. Wang, Design of a distributed energy-efficient clustering algo-rithm for heterogeneous wireless sensor networks. *Computer Communications*, vol. 29, pp. 2230–2237, 2006.

P. Saini and A. K. Sharma, Energy efficient scheme for clustering protocol prolonging the lifetime of heterogeneous wireless sensor networks. *International Journal of Computer Applications*, vol. 6, no. 2, pp. 30–36, 2010.

N. Saxena, A. Roy and J. Shin, Dynamic duty cycle and adaptive contention window based QoS-MAC protocol for wireless multimedia sensor networks. *Computer Networks*, vol. 52, no. 13, pp. 2532–2542, 2008.

Z. Teng and K.-I. Kim, A survey on real-time MAC protocols in wireless sensor networks. *Communications and Network*, vol. 2, no. 2, pp. 104–112, 2010.

J. Tewari, A. Dumka and G. Khan, Sync preempted probability algorithm in the integrated services (IntServ) MPLS network. *International Journal of Science and Research*, vol. 3, no. 6, pp. 696–698, 2014. ISSN (Online): 2319-7064 Impact Factor (2012): 3.358.

R. Tomar, H. Kumar, A. Dumka and A. Anand, Traffic management in MPLS network using GNS simulator using class for different services. In: *2nd International Conference on Computing for Sustainable Global Development (INDIACom)*, 2015.

S. Ullah, B. Shen, S. Riazul Islam, P. Khan, S. Saleem and K. Sup Kwak, A study of MAC protocols for WBANs. *Sensors*, vol. 10, no. 1, pp. 128–145, 2010.

X. B. Wang, Heterogeneous architecture for ad hoc networks. In: *Advanced Materials Research*, vol. 756, Trans Tech Publ. 2013, pp. 1059–1062.

Q. Xuegong, M. Fuchang, Ch. Yan and Y. Weizhao, The protocol of cluster multi-hop transmission based on heterogeneous wireless sensor networks. In: *Proceedings of International Conference on Computational Intelligence and Software Engineering*, 2009, pp. 1–4.

M. A. Yigitel, O. D. Incel and C. Ersoy, QoS-aware MAC protocols for wireless sensor networks: a survey. *Computer Networks*, vol. 55, no. 8, pp. 1982–2004, 2011.

MOBILE WIRELESS SENSOR NETWORKS VI

Chapter 12

Mobility in Wireless Sensor Networks

Since the conception of wireless sensor networks (WSNs), mobility has been in the minds of researchers to be introduced to the domain; this has started becoming a reality nowadays due to recent technological advancements. The notion of mobility in the context of wireless sensor networks has featured the pervasive ubiquitous networks with a variety of applications viz. environmental monitoring, mining, seismic monitoring, health care applications, process monitoring, underwater navigation, inventory tracking, military surveillance, etc. A major variant of wireless sensor networks, when WSNs become equipped with mobility, they are called mobile wireless sensor networks (MWSNs). Moreover, based on the types of resources used in the networks, MWSNs can be further classified into two classes: homogeneous MWSNs and heterogeneous MWSNs. As suggested by the name, homogeneous MWSNs consist of identical sensor nodes, but some or each one is equipped with the ability to move, whereas heterogeneous MWSN consists of a number of mobile sensor nodes which may be dissimilar in functionality and processing/storage/battery powers. An instance of the mobile wireless sensor network is shown in Figure 12.1.

In a general mobile wireless sensor network as shown in Figure 12.1, the static nodes, represented by small circles, are sensing the field to generate the application-specific data, which are in turn being collected by the mobile device. These mobile devices are forwarding this collected data to the nearest sink or access point. Here, the access points take the responsibility of forwarding the sensing information to the base station for end user access.

The advancements in miniaturization technology has made the addition of mobility into WSNs a blessing rather than a problem, as it was being thought of when introduced. A number of studies were conducted on the impact of mobility when introduced in WSNs towards the overall improvements contributed in the

193

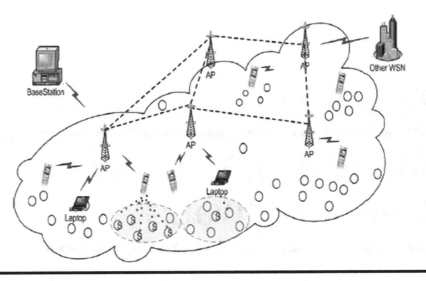

Figure 12.1 Mobile wireless sensor network.

network, and results have confirmed that not only the overall network lifetime is lengthened but also the coverage and data capacity of the network are improved (Liu et al., 2005; Wang et al., 2005; Yarvis et al., 2005). Along with the above cited improvements, MWSNs also deal with delay and network latency. Being a derivative of WSNs, MWSNs' fundamental characteristics are same as that of the WSNs; however, major differences between the two can be listed as follows:

- MWSNs are featured with highly dynamic network topology when compared to WSNs due to mobility added to the nodes.
- Due to highly mobile nodes, existing solutions for medium access and routing etc. in WSNs are not accepted; hence, there is a requirement for the fresh ones in the context of MWSNs.
- Communication links in MWSNs are more prone to breach when compared to WSNs.
- Location estimation is an essential requirement in MWSNs as compared to WSNs.
- MWSNs suffers more with the computational cost in comparison to WSNs.

MWSNs are becoming more and more popular in comparison to wireless sensor networks as they are demonstrating much improved performance over static WSNs. Some of the advantages of mobile wireless sensor networks over non-mobile/static wireless sensor networks can be listed as follows:

- **Coverage:** The most obvious advantage is the enhanced coverage. The nodes enabled with mobility features can definitely cover more area when compared to non-movable nodes in the static WSNs.

- **Quality of Communication:** Due to strategic random deployment of nodes, nodes in the wireless sensor networks (especially in sparse networks) may be located distant from one another and may suffer with the poor quality of communication among them. In MWSNs, mobility of the nodes enables the networks in having better quality of communication as the mobile nodes take on the responsibility of ensuring connectivity when they come into the contact with one another.
- **Reduced Packet Loss:** As can be intuited easily, the higher the number of intermediate hops involved, the higher the chances of packet loss. In MWSNs, due to the presence of mobile nodes, the number of intermediate nodes in data transportation is minimized, which in turn minimizes the chances of data loss due to hopping. Also, along with such minimized data loss, the energy of the nodes is further saved as no more packet retransmission are required.
- **Enhanced Network Lifetime:** WSNs generally suffer with the bottleneck problem in routing as the nodes nearer to the base station are approached again and again by the nodes to transport their data. This results in the early death of the network as the battery of the bottleneck nodes may be depleted because of such repeated exercise due to their unfortunate unalterable locations. In MWSNs, because of the nodes' mobility, energy dissipation is more efficient.

Apart from above mentioned advantages, many more advantages have also been realized viz. high capacity gain etc.

This chapter details the evolution of mobile wireless sensor networks while discussing the main characteristics which discriminate its network design from that of wireless sensor networks. Furthermore, differences in the network architecture in WSNs and that in MWSNs, along with the characterization of mobility in terms of its taxonomy and terminology, will be discussed in the subsequent sections. The chapter concludes with a detailed discussion of mobility patterns and mobility models.

Architecture

A sensor node in a WSN is comprised of basically four units, as shown in Figure 12.2: sensing unit, processing unit, transceiving unit, and power unit; however, the node may also be equipped with some application-dependent additional units viz. location finder, mobilizer, power generator, etc. (Akyildiz et al., 2002) As suggested by the name, the sensing unit is designed to sense application-specific surrounding phenomenon viz. humidity, temperature, pressure, proximity, etc., which is then converted to a digital signal with the help of the adjoining analog-to-digital converter (ADC). The information processed is then handed over to the processing unit for

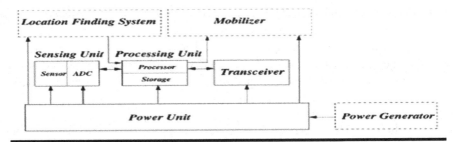

Figure 12.2 Components of a mobile sensor node.

further processing and handling as the processing unit comes together with a small storage unit. The power required for all such activities is supplied by the power unit. The nodes in the MWSN are similar to those in the WSN except for the presence of some additional units to support mobility viz. location finder, mobilizer, and power generator. The functionality of these units can easily be intuited by their names, as the location/position of the node is determined by the location finder unit; the mobilizer unit is responsible for moving the node from one place to another, whereas the power generator is required to fulfill any further energy requirement of the nodes.

In the literature, several network architectures have been proposed since the inception of WSNs. Mainly they can be categorized as in Figure 12.3 (Younis and Fahmy, 2004).

A two-tier network architecture was proposed by Jain et al. (2004) and has been refined in the context of mobile wireless sensor networks (Munir et al., 2007), and it is shown in Figure 12.1.

In the three-tiered network architecture of MWSNs, the sensor nodes are randomly deployed in the sensing field to perform their intended sensing; such sensor nodes form the lowest layer of the network architecture. The nodes equipped with the mobility features form the middle layer whose main responsibility is to collect sensed information from the sensors deployed at the lower layer and to forward the gathered and processed information to the upper layer. These middle layer mobile nodes are also highly computationally capable nodes, along with having extended storage and transmitting capacity, thus forming a heterogeneous version of the sensor networks. The central layer of this architecture, the middle layer, may include a variety of nodes viz. mobile phones, vehicles, people, or even animals, i.e., any mobile agent equipped with data processing capability with regard to the application of interest. Moreover, the most upper layer of the architecture generally refers to the infrastructure network, which can be based on both either wired or wireless technology. Meanwhile, the mobile agents may also collaborate to form an ad hoc network among themselves if required by the application.

More illustratively, the flow of information can easily be intuited with the help of Figure 12.4 (Munir et al., 2007), i.e., the environment-specific sensed data is first collected by the mobile agents when coming into the contact with the sensors at ground

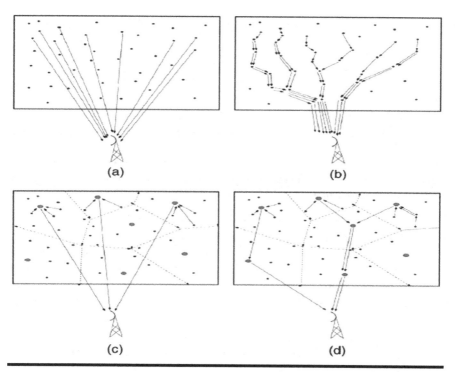

Figure 12.3   Network architectures in static WSN: (a) Single hop without clustering; (b) Multihop without clustering; (c) Single hop with clustering; (d) Multihop with clustering.

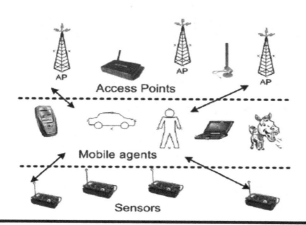

Figure 12.4   Flow of information in the three-tiered sensor network architecture in mobile WSN.

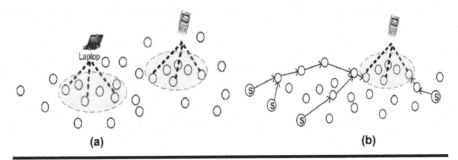

Figure 12.5   Data forwarding modes in mobile WSN: (a) Isolated forwarding; (b) Collaborated forwarding.

zero. Afterwards, the processed information is transported to the access points at the overlay layer destined finally to the base station for end user access.

As in the wireless sensor networks, data sensed by the nodes can be forwarded to the mobile agents in any of the two modes: isolated or collaborated. In the isolated mode of data forwarding, the sensors at the lowest layer keep the originated data after environmental sensing until they encounter the mobile agents. In this mode of data forwarding, data is not communicated to any of the peer nodes, whereas, in the collaborated mode, nodes deployed at the lowest layer may get organized on their own to form an ad hoc network. Data may be communicated to the next node in the topology formed if needed towards the final destination.

Both the modes have their own advantages and disadvantages—in the isolated mode, though nodes' energy is saved as they only have to transmit when they come into the contact of mobile agent, network latency is increased and the network may suffer with data redundancy if nodes are densely deployed. Similarly, in the collaborated mode, the overall energy consumption may be increased, but the quality of data is improved if nodes are facilitated with data aggregation etc. The schemes may be summarized as in Figure 12.5.

Characterization of Mobile WSNs

MWSNs can be characterized by considering a number of aspects listed below (Silva et al., 2014):

1. Mobile elements
2. Movement type
3. Mobility handing

1 and 2 are the physical aspects of the mobility characterization in MWSN whereas 3 is the architectural one. In the following subsections, each of the above characterization will be dealt in detail.

Mobile Elements

Such characterization of mobile wireless sensor nodes is simply based on what element of the network is movable: the sink, i.e., base station(s), or the node(s).

When sinks/base stations are designed to move from one place to another, it is called sink mobility; On its contrary, node mobility refers to the mobility of the sensing components.

a. **Sink mobility:** Furthermore, the sink mobility can be categorized into three classes viz. mobile base station (MBS), mobile data collector (MDC), and rendezvous-based mobility.

Mobile base station refers to the movement of a base station during its operation time to collect the sensory data from the sensing field. A number of works are present in the literature showing that MBS has benefitted the networks in terms of overall network lifetime due to reduced number of hops to reach the destination node and also in terms of improved coverage.

Mobile data collector (MDCs) introduce the concept of high-power nodes called MULEs which enable on-demand data collection from the sensing field. Such nodes, MDCs, can collect the buffered sample whenever they are needed by the network by visiting each sensor node deployed. An MDC can take a variety of paths which are summarized in the following subsection.

Rendezvous-based mobility combines the benefits of MBS and MDC by establishing a few rendezvous points in the network which are closer to the path of mobile agents, whether base station or sinks (Ekici et al., 2006). Sensor nodes in the field are directed to forward their sensed data to these rendezvous points at which data is buffered until retrieved by the mobile collectors (Figure 12.6).

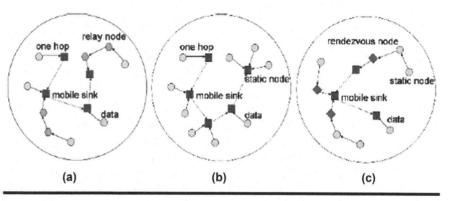

| | | |
| (a) | (b) | (c) |

Figure 12.6   Mobile elements in MWSN: (a) MBS; (b) MDC; (c) Rendezvous.

b. **Node Mobility:** Apart from sink nodes or base station(s), rest of the nodes in the network may also be designed to move as per the requirement imposed by the application, i.e., the nodes may be regularly moving nodes or irregularly; nodes' movement may be intentional or unintentional. Usually the node mobility can be classified into *weak mobility* and *strong mobility* (Sun et al., 2008).

Weak mobility denotes the mobility where change in topology occurs due to unfortunate death of the nodes, mainly because of battery depletion or hardware failure; strong mobility refers to the mobility associated, i.e., nodes are relocated either intrinsically or extrinsically. Furthermore, when nodes are equipped with mobility features, strong mobility is classified as robotic; when they are attached to some moving entity or affected by some external factors viz. wind, water etc., they are classified as parasitic (Dagtas et al., 2007; Cheng et al., 2009).

Movement Type

Mobile wireless sensor networks can also be categorized depending on how the mobile agents are moving in the sensing field, whether sink or the normal sensing nodes. The classification can be achieved as follows:

a. **Random mobility:** As suggested by the name, here the nodes-sink/sensors move randomly within the sensing field.
b. **Predictive mobility:** It represents the pattern where the path of the movements is known a priori along with the speed and other movement details of the moving entity.
c. **Controlled mobility:** In controlled mobility, the nodes' movement is controlled by an external factor.

Mobility Handling

In such characterization, what becomes of concern is the entity taking care of the mobility. In mobile wireless sensor networks, mobility is being handled either at nodes' end or it is supported by the infrastructure; hence, the classification: node-based mobility, network-based mobility, or a hybrid of the two (Silva et al., 2014).

a. **Node-based mobility:** In this class of mobility handling, the nodes themselves manage the mobility in addition to the communication and other application-specific responsibilities. Such burdening comes at the cost of increased complexity, increased energy consumption, and reduced performance.
b. **Network-based mobility:** In this classification, sensors are relaxed from the extra burden of managing mobility, and the network manages the mobility for the nodes either via some external special network nodes or via much

enabled mobile nodes deployed specifically to relieve sensors. Moreover, the additional cost of deploying enabled nodes may be compensated with the improved functioning of the lighter sensor nodes in terms of better performance and enhanced network lifetime.

c. **Hybrid mobility:** This class of mobility handling combines both of the aforementioned approaches as mobility is managed partly by the sensors and partly by the network infrastructure.

Other than the above detailed characterizations, several others are also available in the literature viz. classification based of the protocol layers etc. However, such topics are of more attention and hence they will be discussed in more detail in the upcoming chapters.

Mobility Pattern and Mobility Models

In order to be able to understand the basics of mobility and to implement those in the context of wireless sensor networks, one has to grasp the concept of mobility pattern and mobility model. In the standard literature, sometimes, these two terms are used interchangeably, but these two are completely different concepts. Where mobility patterns refer to the movement pattern of real-life objects viz. people or vehicles, mobility models represent these mobility patterns mathematically (Raja and Su, 2009).

Mobility Patterns

Mobility patterns simply refer to how people and things move in the space. Such patterns can be classified as pedestrian, aerial, vehicular, dynamic medium, robotic, and outer space motion; each being characterized by the properties viz. group behavior, limitations, dimensions, and predictability. Here, *group behavior* denotes if there is any set of nodes staying together for considerably long time; *limitations* refers to the speed and/or acceleration bounds, if any; *dimensions* describes whether the movement is linear or planar or three-dimensional; and, finally, *predictability* implies how well the motion attributes can be specified a priori (Schindelhauer, 2006).

a. **Pedestrian mobility pattern:** The oldest and the most common mobility pattern to walk. It describes the movement pattern of people or animal. In the context of wireless sensor networks, it is materialized by associating sensor nodes to the people or animals moving in their herds. The pedestrian pattern is manifested by two-dimensional, limited speed, and chaotic nature movement which may or may not show group behavior.

b. **Vehicular mobility pattern:** Vehicular mobility pattern is concerned with the movement of trains, cars, bicycles, motorbikes, etc. where sensors are associated to the wheel-based movement which allows high-speed movement.

Vehicles via such mounted/implanted sensors can communicate with one other in order to have much smoother movement. The vehicular movement pattern is characterized by a one-dimensional and high-speed movement showing an extreme group behavior.

c. **Aerial mobility pattern:** Aerial mobility pattern describes the movement of flying objects, e.g., birds, airplanes, etc. which is characterized as two-and-half dimensional (the motion is not completely three-dimensional as the ascent is very costly to the flying object), limited (yet high) speed movement which may or may not show group behavior (Schindelhauer, 2006).

d. **Dynamic medium mobility pattern:** This refers to the event when nodes are moving through a medium such as open air, fluid-surface, or piped gas. The dimension characteristic of this mobility may vary from one-dimensional to three-dimensional depending on the medium of movement. In dynamic medium mobility pattern, group behavior may be shown, but it is unwanted because of data redundancy caused by closely moving sensors (Schindelhauer, 2006).

e. **Outer space mobility pattern:** Outer space mobility pattern describes the movement of space vehicles. Such movement is characterized by limited (yet high) speed, predictable movement showing group behavior (Schindelhauer, 2006).

f. **Robotic Mobility Pattern:** The robotic pattern may represent any of the above mobility scenarios defined by the robot-designer as per the nature of application intended (Schindelhauer, 2006).

Mobility Models

On the contrary to mobility pattern, which deal how people and things move, mobility models deal with the mathematical generalization of the characteristics of mobility patterns. Mobility models play an important role in the simulation or emulation of sensor networks, and they all can be categorized into two broader categories: trace-based and synthetic models. *Trace-based* mobility models refer to the mobility patterns of real-life objects which are kept under thorough observation for a very long period. Though traces provide accurate information about the movement pattern of the objects, capturing the movement trajectory of mobile nodes is quite difficult even when an adequate amount of previous historical data is available from the recurrent mobility patterns. This is exactly where the synthetic models come into picture. *Synthetic models* aim for realistic mapping of the movement pattern of the mobile nodes without any use of traces (Camp et al., 2002).

Synthetic models are further classified into two subcategories: entity and group mobility models, each having further classification as follows:

a. **Entity mobility models**

Entity mobility models describe the scenario where the movement of nodes is independent, i.e., each node can move without any consultation of others.

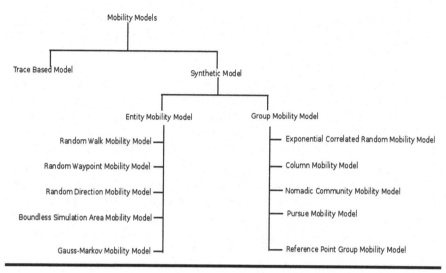

Figure 12.7   Classification of synthetic mobility models.

As detailed in Figure 12.7, entity mobility models can be further classified into a number of classes viz. random walk mobility model, random waypoint mobility model, etc.

i. **Random walk mobility model:** To mimic the erratic movements of most of the objects in nature, the random walk mobility model came into picture. In the random walk mobility model, the mobile agent proceeds in a random direction with a random speed; however, the new speed and the new direction both are chosen from predefined ranges viz. [*minspeed, maxspeed*] and [0, 2Π]. In this model, either after the expiration of a fixed time interval or after the completion of a constant distance, the new distance and the new direction for the agent is computed to be followed by the node. It has already been proved that in the random walk mobility model, the traveling node returns to the origin with a complete certainty, i.e., 1.0 (Weisstein, 1998). The random walk mobility model is the most widely used model and is also referred to as Brownian motion. Moreover, because of the memoryless mobility pattern due to random selection of direction and speed, the model may result in unrealistic movements as the nodes may take sharp directions and may start and stop suddenly (Camp et al., 2002).

ii. **Random waypoint mobility model:** The random waypoint mobility model is a variation of the random walk mobility model in which a pause time is introduced. The mobile agent starts by staying in position for a certain amount of time, which is called pause time. After the expiration of this time, the mobile agent chooses a random direction from the range specified and proceeds with a random speed from a uniformly distributed

204 ■ *A Complete Guide to Wireless Sensor Networks*

interval. Upon arrival at a random destination, the mobile agent takes a pause before restarting the process. The random waypoint mobility model is also widely popular model in the context of concerned network (Camp et al., 2002).

iii. **Random direction mobility model:** A major problem of the random waypoint mobility model is that the nodes start clustering about the center of the simulation area, as it has been proved that the probability of a mobile agent to visit the middle of the simulation area is quite high. In an attempt to alleviate this problem, the random direction mobility model is proposed (Royer et al., 2001). In this model, similar to the random waypoint model, mobile agents take a random direction to move, but they are made to travel up to the boundary of the simulation area. Upon reaching the simulation boundary, nodes take a definite pause time before choosing a random angular direction between 0 to 180 degrees and continuing the process.

iv. **Boundless simulation area mobility model:** This model defines a velocity vector, $\bar{V} = (v, \Theta)$, to represent the nodes' velocity and direction; also, the nodes' location is represented as (x, y). The boundless simulation area mobility model establishes a relationship between the previous and the current values of the mobile agent's direction and speed as per the following set of equations (Haas, 1997):

$$v(t + \Delta t) = \min[\max(v(t) + \Delta v, 0), \text{Vmax}];$$
$$\Theta(T + \Delta t) = \Theta(t) + \Delta \Theta;$$
$$x(t + \Delta t) = x(t) + v(t) * \cos \Theta(t);$$
$$y(t + \Delta t) = y(t) + v(t) * \sin \Theta(t);$$

where $V_{\max}$ is the maximum defined velocity, Δv is the change in velocity which is uniformly distributed between $\left[-A_{\max} * \Delta t, A_{\max} * \Delta t \right]$, $A_{\max}$ is the maximum acceleration of the concerned mobile agent, $\Delta \Theta$ is the change in direction which is uniformly distributed between $[-\alpha * \Delta t, \alpha * \Delta t]$, and α is the maximum angular change in the direction of node's movement.

Moreover, the simulation boundary handled in the boundless simulation area mobility model is quite different from rest of the models. In this model, once the nodes reach the boundary, instead of reflecting off or stopping, they continue to travel and reappear on the opposite side of the simulation, hence producing a torus-shaped area.

v. **Gauss–Markov mobility model:** The Gauss–Markov model has been designed to adapt to the different levels of randomness via one tuning parameter (Camp et al., 2002). In this model, mobile agents are assigned some initial values of speed and direction which are successively updated at fixed intervals of time as per the following set of equations (Liang and Haas, 1999; Tolety, 1999):

$$s_n = \alpha\, s_{n-1} + (1-\alpha)\overline{s} + \sqrt{\left(1-\alpha^2\right)} s_{x_{n-1}}$$

$$d_n = \alpha\, d_{n-1} + (1-\alpha)\overline{d} + \sqrt{\left(1-\alpha^2\right)} d_{x_{n-1}}$$

where s_n and d_n are the new speed and direction of the mobile agent at the time interval n; α, where $0 \leqslant \alpha \leqslant 1$ is the tuning parameter used to vary the randomness; here, when $\alpha = 0$, total random values (Brownian motion) is obtained and when $\alpha = 1$, linear motion is obtained. $\overline{s}$ and $\overline{d}$ are mean values of speed and direction respectively as $n \to \infty$; and $s_{x_{n-1}}$, and $d_{x_{n-1}}$ are the random variables from Gaussian distribution.

Now, with the help of the current values of location, speed, and direction, the next location of the nodes (at time interval n) can be calculated as follows:

$$x_n = x_{n-1} + s_{n-1} \cos d_{n-1}$$

$$y_n = y_{n-1} + s_{n-1} \sin d_{n-1}$$

where $\left(x_n, y_n\right)$ and $\left(x_{n-1}, y_{n-1}\right)$ are positions of the mobile agent at nth and n-1th time intervals respectively and s_{n-1} and d_{n-1} are the speed and direction at n-1th time interval respectively.

b. **Group mobility models**

Contrary to the entity mobility models, group mobility models describe the scenario where nodes' movements are dependent on one another, i.e., movement occurs in consultation with others because in many applications of MWSN, nodes may be required to collaborate with each other to achieve the objective viz. military applications etc.

As detailed in Figure 12.7, group mobility models can be further classified into a number of classes viz. exponential correlated random mobility model, column mobility model, etc.

i. **Exponential correlated random mobility model:** This model defines a motion function to create nodes' movements as per the following equation (Hong et al., 1999):

$$b(t+1) = b(t)e^{-1/\tau} + \left(\sigma \sqrt{1 - \left(e^{-\frac{1}{\tau}}\right)} \right) r$$

where $\vec{b}(t)$ is the position of the nodes at time t and $\vec{b}(t+1)$ is that at time (t+1); τ is the rate of change from node's previous position to current one and r is the random Gaussian variable with variance σ.

ii. **Column mobility model:** This mobility model has proved its significance in the design of scanning and searching applications (Sanchez and Manzoni, 2001). In this model, a number of mobile agents move together around an axis (or column) which itself moves in a forward direction. To implement the idea, an initial reference grid is defined (Sanchez, 2002) and each node is given a reference point in this reference grid around which the node can move as per some entity model. Moreover, the new reference points from the moving nodes is defined as follows:

$$Reference_{New} = Reference_{old} + Advance - vector$$

where $Reference_{New}$ and $Reference_{old}$ are the new and previous reference points of a node respectively and the Advance–vector is a predefined offset computed as per some random distance and random angle between 0 and moving the reference grid.

As in Figure 12.8, black dots represent the reference points for their respective nodes, represented using small circles. The model can be easily intuited by the readers.

iii. **Nomadic community mobility model:** This model imitates the movement of an ancient nomadic group of people. In contrast to the column mobility model, all the mobile agents in the group share a common reference point around which they move as per some entity mobility model. When the reference point changes, then all the nodes move to the new area and start roaming around the reference point (Sanchez and Manzoni, 2001; Sanchez, 2002) (Figure 12.9).

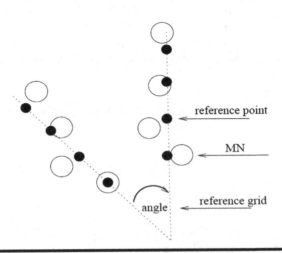

Figure 12.8 Movement of four mobile agents using column mobility model.

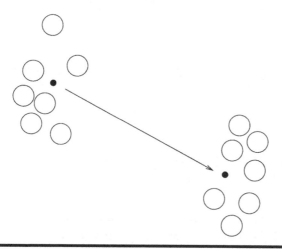

Figure 12.9 **Movement of four mobile agents using nomadic community mobility model.**

iv. **Pursue mobility model:** As suggested by the name, this aims to model the mobile agents pursuing a particular target. The new position of the nodes can be obtained as per the following equation (Camp et al., 2002; Sanchez and Manzoni, 2001; Sanchez, 2002):

$$New\_Position = Old\_Position + Acceleration\left(target - Old\_Position\right)$$
$$+ Random\_Vector$$

where *Acceleration(target-Old_Position)* is the information on the movement of the node being pursued and *Random_Vector* is the common random offset obtained via some entity mobility model for each of the mobile agents.

Here, in Figure 12.10, the white node is the one being pursued, and black ones are the pursuing nodes.

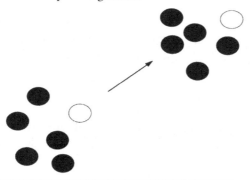

Figure 12.10 **Movement of six nodes under the pursue mobility model.**

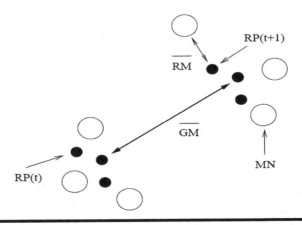

Figure 12.11 Three nodes' movement as per the RPGM.

v. **Reference point group mobility model:** Reference point group mobility model (RPGM) not only represents the random motion of the group of mobile agents, but also describes that of the individual nodes in the group. With each node in the group, there comes a reference point which collectively forms a logical center for the group. Group movement is defined on the basis of traveling path of this logical center whose movement can be characterized by random waypoint mobility model. Moreover, motion of the individual node in the group can also be implemented as the random waypoint mobility model but without any notion of pause time while the group is moving. It is only when the group reference point reaches a destination that all the member nodes, i.e., mobile agents, may pause for the same amount of time. RPGM has been widely accepted for the applications viz. rescue operations, emergency crews, military divisions/platoons, group tours, etc.

In Figure 12.11, the black ones are the reference points of the mobile agents being represented by the white ones. RP(t) and RP(t+1) describes the reference points at t and t+1 time instances. $\overrightarrow{GM}$ refers to the group motion vector for the calculation of logical center of the group. $\overrightarrow{RM}$ is the random motion vector which is combined with the updated reference point to represent the random motion of each node about its own reference point.

Moreover, selection of an appropriate and application-specific mobility model for the network always has a significant effect on the simulation and performance evaluation of the network.

References

I. F. Akyildiz, W. Su, Y. Sanakarasubramaniam and E. Cayirci, Wireless sensor networks: A survey. *Computer Networks*, vol. 38, no. 4, pp. 393–422, 2002.

T. Camp, J. Boleng and V. Davies, A survey on mobility models for ad hoc network research. *Wireless Communications and Mobile Computing*, vol. 2, pp. 483–502, 2002.

M. Cheng, M. Kanai-Pak, N. Kuwahara, H. I. Ozaku, K. Kogure and J. Ota, Dynamic scheduling based inpatient nursing support: Applicability evaluation by laboratory experiments. In: *Proceedings of the 3rd ACM International Workshop on Context-Awareness for Self-Managing Systems*, 2009, pp. 48–54.

S. Dagtas, Y. Natchetoi and H. Wu, An integrated wireless sensing and mobile processing architecture for assisted living and healthcare applications. In: *Proceedings of the 1st ACM SIGMOBILE International Workshop on Systems and Networking Support for Healthcare and Assisted Living Environments*, 2007, pp. 70–72.

E. Ekici, Y. Gu and D. Bozdag, Mobility-based communication in wireless sensor networks. *IEEE in Communications Magazine*, vol. 44, pp. 56–62, 2006.

Z. Haas, A new routing protocol for reconfigurable wireless networks. In: *Proceedings of the IEEE International Conference on Universal Personal Communications (ICUPC)*, 1997, pp. 562–565.

X. Hong, M. Gerla, G. Pei and C. Chiang, A group mobility model for ad hoc wireless networks. In: *Proceedings of the ACM International Workshop on Modeling and Simulation of Wireless and Mobile Systems (MSWiM)*, 1999.

S. Jain, R. C. Shah, W. Brunette, G. Borriello and S. Roy, Exploiting mobility for energy-efficient data collection in sensor networks. In: *Proceedings of the IEEE Workshop on Modeling and Optimization in Mobile, Ad Hoc and Wireless Networks*, WiOpt ,2004.

B. Liang and Z. Haas, Predictive distance-based mobility management for PCS networks. In: *Proceedings of the Joint Conference of the IEEE Computer and Communications Societies (INFOCOM)*, March 1999.

B. Liu, P. Brass, O. Dousse, P. Nain and D. Towsley, Mobility improve coverage of sensor networks. In: *Proceedings of ACM MobiHoc*, 2005.

S. A. Munir, B. Ren, W. Jian, B. Wang, D. Xie and J. Ma, Mobile wireless sensor network: Architecture and enabling technologies for ubiquitous computing. In: *Proceedings of 21st International Conference on Advanced Information Networking and Applications Workshops (AINAW'07)*, 2007.

A. Raja and X. Su, "Mobility handling in mac for wireless ad hoc networks," *Wireless Communications and Mobile Computing*, vol. 9, pp. 303–311, 2009.

E. Royer, P. M. Melliar-Smith and L. Moser, An analysis of the optimum node density for ad hoc mobile networks. In: *Proceedings of the IEEE International Conference on Communications (ICC)*, 2001.

M. Sanchez, Mobility models & lt. Available at: http://www.disca.upv.es/misan/mobmo del.htm&gt, accessed on May 30 2002.

M. Sanchez and P. Manzoni, Anejos: A java based simulator for ad-hoc networks. *Future Generation Computer Systems*, vol. 17, no. 5, pp. 573–583, 2001.

C. Schindelhauer, Mobility in wireless networks. In: *SOFSEM*, 2006, pp. 100–116.

R. Silva, J. S. Silva and F. Boavida, Mobility in wireless sensor networks–survey and proposal. *Computer Communications*, vol. 52, pp. 1–20, 2014.

210 ■ *A Complete Guide to Wireless Sensor Networks*

Y. Sun, O. Gurewitz and D. B. Johnson, RI-MAC: A receiver-initiated asynchronous duty cycle mac protocol for dynamic traffic loads in wireless sensor networks. In: *Proceedings of the 6th ACM Conference on Embedded Network Sensor Systems* 2008, pp. 1–14.

V. Tolety, Load reduction in ad hoc networks using mobile servers. Master's thesis, Colorado School of Mines, 1999.

W. Wang, V. Srinivasan and K.-C. Chua, Using mobile relays to prolong the lifetime of wireless sensor networks. In: *MobiCom*, 2005.

E. W. Weisstein, *The CRC Concise Encyclopedia of Mathematics*. CRC Press, Boca Raton, FL, 1998.

M. Yarvis, N. Kushalnagar, H. Singh, A. Rangrajan, Y. Liu and S. Singh, Exploiting heterogeneity in sensor networks. In: *Proceedings of IEEE INFOCOM* 2005.

O. Younis and S. Fahmy, HEED: A hybrid energy-efficient distributed clustering approach for ad hoc sensor networks. *IEEE Transaction on Mobile Computing*, vol. 3, no. 4, pp. 660–669, 2004.

Chapter 13

Localization in Wireless Sensor Networks

Introduction

Localization is one of the most important aspects when considering the mobility in wireless sensor networks (WSNs) and mobile wireless sensor networks (MWSNs). This chapter focuses on issues of localization and discusses various localization protocols of WSNs and WMSNs.

With an increase in usage of MWSNs, the need for mobilized nodes which can sense or monitor cooperatively the environmental and physical conditions also increases. The mobility within the network can be achieved by means of equipped sensors with mobilizers that enable them to change the locations. Mobility can also be achieved by attaching the sensors with transporters such as robots or vehicles. Mobility can be given to the base station (BS) which acts as a sink for collecting information from static sensors, or mobility can be provided to sensors which collect the environmental parameters and transmit information to the base station in a regular fashion, or nodes can serve as data-relaying nodes within the network when recipient is not within the range of transmitters.

MWSNs can be categorized at the node level into the following categories based on their roles as:

Mobile embedded sensors: These are sensors which don't control their own movement; instead, their movement is controlled by other external factors. Such sensors are attached to some moving object which moves as per the moving objects.

Mobile actuated sensors: These sensors have locomotive capabilities which make them move around the sensing region. This mobility is done in a controlled manner which can cover a larger area and follow specific phenomena.

Data mules: These sensor nodes are not mobile in nature; instead, they require a mobile device to collect data and forward it to the base station. Data mules can automatically recharge their power sources.

Access point: Nodes of the network works as network access points when a node drops off the network and mobile nodes position themselves to maintain the network connectivity, and these nodes works as access points.

MWSNs can be categorized into flat, two-tier and three-tier hierarchical architecture. Flat architecture consists of a set of heterogeneous devices that communicate in an ad hoc manner. Flat architecture can use static or mobile nodes which communicate with each other within the same network. Two-tier architecture consists of set of static and mobile sensor nodes. Mobile nodes are used to create an overlay network or act as data mules that help in the moving of data within the network. Overlay networks consist of mobile devices with high processing power, longer communication ranges, and high bandwidth. The overlay network is used to make connection between nodes. Three-tier architecture consists of a set of static nodes, mobile nodes, and access points. Stationary nodes pass the data to mobile nodes, which forward the data to access points, which are further used for setting up the communication. Three-tier architecture is used to cover a wider area and is also compatible with a further number of applications.

Localization is the estimation of the position and spatial coordinates of nodes and hence is an important factor in the mobility of nodes. Localization helps the nodes to be aware about their accurate position with respect to the entire network. Localization of nodes in WMSNs can be divided into three sub-parts as: *coordination phase, measurement phase*, and *localization phase*. In the coordinate phase, sensor nodes are initialized for the process of localization and includes synchronization of clock and notification that localization process is about to begin. In the measurement phase, each sensor node within the network transmits a signal which is received by receiver node to calculate the measurement of distance and number of hops from source to receiver. This technique can be further classified into two as: range based and range free. The localization phase involves the use of different optimization techniques or computation methods meant to find the location and position of sensor nodes.

Localization is initiated by coordination among groups of nodes; one or two nodes among these nodes emit a signal and some property of this signal is observed by one or more receivers and transformation of signal measurement into position is done by means of localization algorithms.

Classification of Localization in WMSNs

There are different mobilization techniques and algorithms used for finding the location of sensor nodes within the network that differ from application to application. In the subsequent chapters, we will focus on all such types of mobility techniques and algorithms proposed by different research scholars in this field.

Classification of Localization Protocol in MWSNs.
Within MWSN, localization protocol can be based on following factors:

1. Estimating location
2. Cooperative or anchor based
3. The state of information, which can be symmetric or asymmetric and can be determined by radio transmission between anchor nodes and unknown nodes

Taking view of the mobility of nodes, three different scenarios can be classified:

1. Mobile sensors, static anchors
2. Static sensors, mobile anchors
3. Mobile sensors, mobile anchors

A brief overview of classification of localization protocol is shown in Figure 13.1.

Localization protocol can be classified into centralized- or distributed-based localization algorithms.

Localization techniques can be categorized as mentioned above into range-based and range-free techniques. For calculation of location, range-based techniques use distance or angle as parameters. There are numerous range-based techniques which use special hardware for determining estimated time of arrival (ToA), received signal strength (RSS), direction of angle (DoA), angle of arrival (AoA), time difference of arrival (TDoA), and frequency difference of arrival (FDoA). In terms of accuracy, range-based techniques are more accurate than range-free techniques. On the other hand,

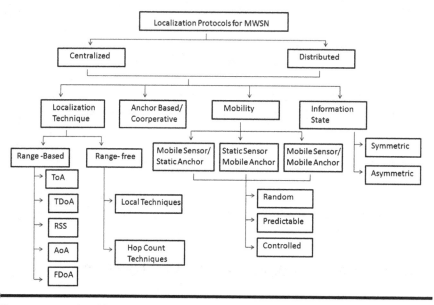

Figure 13.1   Classification of localization protocol in WMSN.

range-based techniques require extra hardware for ToA, RSS, DoA, AoA, TDoA, and FDoA; they also require more cost, noise sensitivity, complexity, and energy consumption than range-free technique. Range-free techniques can be further classified into two main classes as local technique and hop-counting technique, where local technique is based on high anchor density, which enable the nodes of the network to hear several anchors, whereas hop counting technique is based on the mechanism of flooding.

Another way of classifying localization protocols is based on anchor-based or cooperative-based. Anchor-based localization protocol uses one or more anchor nodes (which can be devices like robots, vehicles, etc.) which use techniques like GPS to have prior knowledge of their geographic location. These anchor nodes have sufficient resources, are capable of making local decisions, and can execute actions as per the inputs received from sensor nodes. These sensor nodes can adjust their power of transmission so as to transmit their beacon signal at different levels of power. Other sensor nodes of the network can determine their location by using the location of their nearby anchor nodes. The localization accuracy on anchor-based localization protocol depends upon distribution and number of anchor nodes. Anchor nodes can be used to obtain global coordinates. Anchor nodes are expensive in nature as they use GPS for finding the positioning of nodes, and, also, these nodes cannot be used for indoor purposes due to issue of line-of-sight communication. Cooperative localization is also termed as "anchor-free" localization, as in this localization algorithm, the location of nodes is estimated by means of cooperation among sensor nodes, thus forming a local coordinate system to determine the location of nodes in a relative manner. This protocol is economical as it does not use GPS technology for finding the location of sensor.

The mobility-based classification of WMSNs can be further classified into three major categories as: random mobility, predictable mobility, and controlled mobility. In random mobility, the nodes are mobile or move randomly and freely within the area of deployment without any restrictions. The destination, direction, and speed of such nodes are decided in a random manner and independently. In predetermined mobility, the nodes are mobile or move within the network in a predetermined path. In this category, the motion of nodes is known but can't be changed. In controlled mobility, the mobility devices moving to a specific destination follow a defined mobility pattern for common objectives like localization and exploration.

Classification of localization protocol can be done based on the state of information of data which can be symmetric or asymmetric.

In recent years, researchers have proposed for combined range-based techniques termed as hybrid positioning which exhibits sufficient coverage and accuracy.

Review Work by Researchers in This Direction

There are many researchers who have contributed localization techniques and algorithms for WMSNs. This contribution of algorithms is divided into centralized- or distributed-based. In the centralized approach, the location of nodes is determined by the sink node, which is the central point of the network. In the centralized approach, the sensor nodes collect information of their environmental variables and transmit the information to the sink, which analyzes, calculates, and transmits the position to sensor nodes. As far as energy dissipation is concerned, the centralized approach is not good as it consumes more energy in moving data back to the base station. The centralized approach is also not suitable for scalable network, as when the number of nodes increases, the centralized algorithm takes longer to converge and also the energy dissipation in convergence is more which decreases the performance of the algorithm. Due to the single point of action in the centralized approach, the problem of network failure is higher as all functioning in this approach depends upon the sink node.

The distributed approach uses all nodes within the network for finding the path and thus distributes the working of localization within the network. In this approach, all nodes communicate with their neighbors and exchange information of neighboring to estimate distance and position of the nodes. In terms of energy consumption, the distributed approach is better than the centralized, but, being distributed in nature, the accuracy of location of nodes is less in distributed than in centralized. The distributed approach for scalable network is better than centralized as the work is distributed among nodes in this approach.

Amundson and Koutsoukos (2009), as well as Khelifi et al. (2005), in their papers, presented an overview of localization algorithm for WMSNs. They discuss architecture, the advantages of mobility, and compare WMSNs with WSNs. They also discuss forms of mobility and their strategies. Khelifi classified the localization algorithm into event-driven and time-driven. An event-driven localization algorithm is used when an interesting event occurs, such as monitoring a river flowing through dense forest. Thus, the event-driven approach is used to provide nodes' location in timely manner. On the other hand, the time-driven localization algorithm is used to compute the real-time location of sensor nodes by repeated execution of localization algorithm.

Han et al. (2016) in his paper presented a survey on mobile anchor node assistant localization algorithm (MANAL). In his paper, the author categorizes MANAL into two categories as based on the mobility model and based on the path planning scheme. In the mobility model scheme, pre-existing mobility models are used for the movement of anchor nodes without considering network parameters and localization conditions. Whereas in the path planning scheme, the anchor nodes are moved based on path planning scheme which are designed for WSN localization. The author in his paper categorizes the mobility model as individual mobility model and group mobility model, where the individual mobility model is further classified into three sub-groups as based on random walk model, based on random waypoint (RWP) model, and based on random directional model; the group model is further classified into synthetic mobility model and hybrid mobility model as shown in Figure 13.2.

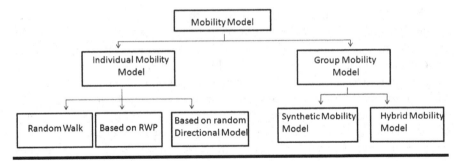

Figure 13.2   Categorization of mobility model as per Han et al. (2016).

The path planning scheme can be of two types: static or dynamic. The localization based on path planning scheme can provide a high localization ratio and accuracy and consumes less energy as compared to mobility model.

In Khelifi et al. (2005), the authors have reclassified mobile localization algorithm based on their triggering nature. The authors mention some event-driven and time-driven approaches to address localization issues based on their triggering nature. The authors have focused on different mobile strategies and mobile localization algorithms.

Classification of Localization Algorithms

Mobile sensors and static anchors: Mobile sensors and static anchors depend on various parameters, such as localization technique (LT), localization accuracy (LA), energy consumption (EC), communication cost (CC), computational complexity (CComp), robustness and transmission range (TR). Localization techniques are based on range-based and range-free schemes. Localization accuracy depends on the frequency of location updates which is high and harsh and noisy conditions. Energy consumption of the localization approach is efficient. The communication cost of localization is not low enough, whereas the computational complexities of localization are acceptable. Localization has robustness against the phenomena of diffraction, scattering, shading, multipath propagation, miscellaneous noise sources, and harsh factor conditions. The transmission range of localization can be of symmetric or asymmetric. One of the factors that may affect localization is node/anchor density. The accuracy of localization increases with the increase in number of anchor nodes as nodes receive more location announcements in dense anchor nodes.

This is a technique that can be illustrated by using a real-time example of military application, where military bases are anchors and soldiers are located in the function of their positions. The algorithm used under this can be classified into:

1. Received information-based localization algorithm
2. Artificial neural network-based localization algorithm

Received information-based localization algorithm: This algorithm is based on idea of determining the coordinates of mobile sensor nodes from the information gathered from anchor nodes. Jabbar et al. (2010) proposed a novel approach, power tunng anchors (PTA), for providing localization within the network. In this approach, anchor nodes, on receiving a request from the base station, broadcast beacon signal with their maximum level of power. The mobile sensor nodes on hearing the query elect three neighboring anchor nodes which all together form a triangle. The mobile nodes acknowledge to these three anchors by retransmitting beacons with reduced power. Anchor nodes on receiving the acknowledgement reduce power level by one step and retransmit the beacon signals. The process is repeated till no request is received by any selected anchor. This approach increases the accuracy of localization with an increased percentage of node localization and also increases the range of radio and the number of anchors deployed within the network.

Event-triggered algorithm is proposed by Singh and Saini (2015) and is based on infrared (IR) fingerprint and received signal strength indicator (RSSI) techniques. Upon receiving messages from neighboring anchor nodes, the anchor nodes record the receiving power and coordinates of each neighboring node. This algorithm creates a database of each fingerprint created and the power levels of signal received from each anchor node. It records the RSSI and coordinates value of each received anchor node and thus measures the distance. The algorithm results in a low average localization error with a large number of anchor nodes.

Wang and Han (2011) proposed a new algorithm, self-adapting localization for mobile nodes (SALMN), which is self-adapting and dynamic in nature, in order to overcome the problem of localization of mobile nodes. This algorithm is used to find localization only when two or one beacon node are within the range of communication and is achieved by the creation of motion model. This algorithm achieves better accuracy in finding the position of nodes and achieves a high rate of localization and also extends the lifetime of network.

Artificial neural network-based localization algorithm: The artificial neural network approach is used to moderate the effect of miscellaneous conditions and noise sources. Gholami et al. (2013) proposed an approach for the harsh and noisy conditions of manufacturing environments. This algorithm uses ultrasound for estimating the distance between two sensors, and this approach can be affected by factors like noise sources, ambient conditions, and physical obstructions.

Mobile anchor and static nodes: This approach can be used in applications like monitoring of forest fires, where the animals are anchors and sensors are dropped in the forest. This approach can be categorized into further two categories as:

1. Measurement confidence-based localization algorithm
2. Geometric localization algorithm

Measurement confidence-based localization algorithm: This algorithm assigns a confidence value to a distance measurement where the localization algorithm intelligently localizes each sensor based on the confidence distance, which gives improved location accuracy.

Ding et al. (2010) proposed an algorithm using this approach termed as virtual ruler, which is based on mobile beacons. In this approach, a pair of beacons are attached to a vehicle which moves around a pair of sensors to observe different values of distance measured between the same pair of sensors from different perspectives. The proposed approach achieves a long distance measurement without violation of energy constraint.

Conclusion

This chapter discussed localization in wireless sensor networks and summarized the localization algorithm used in mobile wireless sensor networks and how they are differentiated in different parameters. This chapter also presented different protocols proposed by different researchers in the field of localization.

References

I. Amundson and X. D. Koutsoukos, A survey on localization for mobile wireless sensor networks. In: *Mobile Entity Localization and Tracking in GPS-less*. Springer, Berlin, Heidelberg, 2009, pp. 235–254.

Y. Ding, C. Wang and L. Xiao, Using mobile beacons to locate sensors in obstructed environments. *Journal of Parallel and Distributed Computing*, vol. 70, no. 6, pp. 644–656, 2010.

M. Gholami, N. Cai and R. W. Brennan, An artificial neural network approach to the problem of wireless sensors network localization. *Robotics and Computer-Integrated Manufacturing*, vol. 29, no. 1, pp. 96–109, 2013.

G. Han, J. Jiang, C. Zhang, T. Duong, M. Guizani and G. Karagiannidis, A survey on mobile anchor node assisted localization in wireless sensor networks. *IEEE Communications Surveys and Tutorials*, vol. 18, 2016.

S. Jabbar, M. Z. Aziz, A. A. Minhas and D. Hussain, A novel power tuning anchors localization algorithm for mobile wireless sensor nodes. In: *Proceedings of the 10th IEEE International Conference on Computer and Information Technology*, 2010, pp. 2441–2446.

M. Khelifi, I. Benyahia, S. Moussaoui and F. Naït-Abdesselam, An overview of localization algorithms in mobile wireless sensor networks. In: *Proceedings of IEEE International Conference on Protocol Engineering (ICPE) and International Conference on New Technologies of Distributed Systems (NTDS)*, 2005, pp. 1–6.

P. Singh and H. S. Saini, Average localization accuracy in mobile wireless sensor networks. *Journal of Mobile Systems, Applications and Services*, vol. 1, no. 2, pp. 77–81, 2015.

K. F. Ssu, C. H. Ou and H. C. Jiau, Localization with mobile anchor points in wireless sensor networks. *IEEE Transactions on Vehicular Technology*, vol. 54, no. 3, pp. 1187–1197, 2005.

J. Wang and T. Han, A self-adapting dynamic localization algorithm for mobile nodes in wireless sensor networks. *Procedia Environmental Sciences*, vol. 11, pp. 270–274, 2011.

Chapter 14

Coverage and Connectivity in Mobile Wireless Sensor Networks

Wireless sensor networks (WSNs) feature a number of nodes whose task is to sense the environment and to forward the sensed information to a central processing body called the base station or sink. The base station further processes the received information and allows end users to access the contents if provisioned by the application. A wireless sensor network can be classified into a number of ways according to the *types* of nodes involved, or *deployment strategy* being followed, or *network architecture* under consideration, or *mobility* involved in the network, etc. In fact, there are numerous ways to classify the wireless sensor networks in addition to the ones mentioned above.

When the type of node involved is taken as the classification criteria, WSNs can be categorized as static wireless sensor networks (i.e., SWSNs), mobile wireless sensor network (i.e., MWSNs; also known as WMSNs, i.e., wireless mobile sensor networks), hybrid wireless sensor networks (HWSNs), and wireless sensor and robot networks (i.e., WSRNs) based on whether the type of nodes in the networks are *static, mobile, both static and mobile,* and *mobile robots*, respectively. Here, SWSNs are comprised of static nodes only and are the traditional wireless sensor networks; MWSNs involve only mobile nodes which can change their location immediately after the deployment in the sensing field; HWSNs consists of both types of the nodes—static as well as mobile—and is more a popular one; and lastly, WSRNs are composed of mobile robots which are used to carry sensor nodes from one place to another in order to satisfy the needs of application. In the WSRNs, nodes once dropped by the mobile robots stay static for the remaining course of network actions (Mei et al., 2007), i.e., a WSRN comprises only mobile robots and static sensor nodes and no such other nodes with additional movement ability are there in the network.

When deployment strategy is taken as the network classification criteria, WSNs are said to be either *random* or *deterministic*. In the random WSNs, nodes are deployed randomly whereas in the deterministic WSNs, nodes are placed at planned locations as per the intended applications. However, random WSNs are more popular when compared with deterministic WSNs.

With regard to the network-architectures, WSNs are majorly categorized as the ones with *flat network architecture* and with *hierarchical network architecture*. In the flat architecture WSNs, every node in the network has peer capacity and can play any role viz. originators or forwarders, whereas in the hierarchical WSNs, some nodes are chosen over others to perform heavy duty operations as compared to the others.

Mobility can be controllable and uncontrollable. Uncontrollable mobility can create a more complex network to work with, whereas the controllable mobility can be treated as an additional advantages to facilitate the network operations.

WSN has been widely accepted as a very popular standard for a variety of applications, ranging from monitoring (a particular field of interest) to tracking (an object of interest). Such a vast range of applications result in a number of design, operational, and managerial challenges in WSNs. Among the challenges provoked, coverage is a very significant issue which requires one or more points/regions to be observed and/or monitored as per the application requirements.

In addition to the ones mentioned above, there are numerous other ways to classify the wireless sensor networks; one such way is the coverage anticipated by the application of interest. All the nodes collaboratively attempt to maintain the required coverage. However, the deployment of nodes affects the coverage a lot. When the nodes are deployed deterministically, nodes can be easily appointed to monitor the field of interest (FoI), but, in case deployment is random, achieving the desired coverage is more complicated. From the coverage viewpoint, WSNs can be categorized as 1-covered or k-covered depending on the number of sensor nodes (one or more) employed to cover the intended area (i.e., partial coverage) or the entire network region (i.e., full area coverage).

Moreover, coverage alone is not sufficient to fulfill the requirements of application if nodes are not able to communicate their data to the sink, i.e., not only the coverage preservation is required, but also maintenance of the network connectivity is equally required for the success of network. Coverage and connectivity both complement each other in such a way that one is meaningless in the absence of another. More illustratively, when the network is fully covered but nodes are not able to communicate to the base station, as they might fail to locate any path to the sink, the overall network operation succumbs. Similarly, even in the presence of fully connected network, if the nodes are losing coverage in some particular areas, the system is said to be partially functioning.

In this chapter, the focus is on the issue of coverage and connectivity in the context of MWSNs. The chapter provides an in-depth discussion on the basic notions of coverage and connectivity in wireless sensor networks featured with mobility.

Nomenclature

A number of preliminary concepts are required in dealing with the coverage and connectivity in the context of wireless (mobile) sensor networks as follows.

Static Nodes

Nodes which are able to perform the operations, viz. sensing, transmitting, receiving, and processing etc., but are not able to move from one place to another are called *static nodes*.

Mobile Nodes

Along with the abilities of static nodes, mobile nodes are also equipped with the facility allowing them to move from one place to another. Mobile nodes are the significant ones when consideration is being given to coverage and connectivity as they can be used to fill the coverage gaps due to the failures of battery depletion of some existing nodes in the network.

k-Coverage

A point or a region in the network is said to be k-covered if being covered by k-number of sensor nodes. The advantage of k-coverage is that the point or portion of interest remains under observation even if (k-1) nodes have failed for any reason, viz. battery depletion, sudden fault due to surrounding environment, etc.

k-Connectivity

A network is said to be k-connected if between every pair of communicating nodes there exists k-disjoint paths, i.e., even the failure of (k-1) path could not down the network (Liu et al., 2005).

Full/Blanket Coverage

Full coverage or blanket coverage requires each point in the field of interest to be brought under observation using 1 or more number of nodes.

Target Coverage

On contrary to the full or blanket coverage, target coverage requires only a few points (of interest) to be covered or brought into observation. Target coverage is basically achieved by employing deterministic deployment of nodes.

Path/Barrier Coverage

Path coverage is treated as a special case of target coverage in which all the points (of interest) to be monitored are put over a definite path. The most suitable application for such type of coverage is intruder detection in which security breaches can be easily detected.

Static Coverage

Static coverage corresponds to static nodes deployed densely in order to provide comprehensive coverage in the field of interest. Since nodes are static and cannot move, any node failure may lead to a coverage hole if the network is just 1-covered (i.e., covered by just one node).

Dynamic Coverage

The coverage attained by the mobile nodes is termed as dynamic coverage. As the nodes are mobile, they can be easily motivated to fill in the coverage gaps due to any node's failure. They can also be used to improve the overall network coverage and efficiency of the network by provisioning the nodes' placement at appropriate places.

Coverage Degree

Coverage degree refers to the number of nodes monitoring the point (of interest), object (of interest), or area (of interest). A point/object/area is said to be covered with degree k, if k number of sensors are consistently monitoring that.

Coverage Determination

Coverage determination is the process of determining the degree of coverage.

Convergence Time

Convergence time refers to the time taken by a network in getting stable and functioning by following some self-organizing algorithm (Li et al., 2009).

Coverage Deployment

Coverage deployment refers to the minimal number of nodes to be deployed in the sensing field in such a way that meets the requirement of application, especially in attaining the anticipated coverage degree.

Sensing Range

Sensing range represents the area which can be sensed by a sensor node. It is denoted by R_s.

Communication Range

Communication range represents the area up to which a node can transmit. It is denoted by R_c.

Sensing Model

As stated earlier, a wireless sensor network consists of a number of nodes, each capable of sensing, processing, and transceiving due to the respective functional units.

Let a WSN be comprised of n number of nodes such as $s_1, s_2, s_3, \ldots, s_n$.

In view of representing a WSN using graph theory, i.e., *G(S,E)*, let S is the set of all the nodes deployed in the network and E be the set of edges representing links between a pair of nodes as follows:

$$S = \left\{ s_1, s_2, s_3 \ldots, s_n \text{ such that } s_i \in S \right\}$$

$$E = \left\{ \left(s_i, s_j \right) \in S \times S \mid d \left(s_i, s_j \right) \leq R_c \bullet s_i, s_j \in S, i! = j \right\}$$

Here, $d(s_i, s_j)$ is the Euclidean distance between the sensors s_i and s_j.

Let a sensor s_i is deployed at a point (x_i, y_i), then the Euclidean distance of this node s_i from the point P (x,y) is denoted by $d(s_i, P)$ as follows:

$$d\left(s_i, P \right) = \sqrt{\left(x_i - x \right)^2 + \left(y_i - y \right)^2}$$

There are two sensing models for the evaluation of wireless sensor network: disc sensing model and irregular sensing model.

Disc Sensing Model

The disc sensing model is based on simple physical phenomenal observation in which sensing is represented as an event of detection if the object is found to be in the sensing range of the node as follows:

$$p\left(s_i, P \right) = \begin{cases} 1, & d\left(s_i, P \right) \leq R_s \\ 0, & d\left(s_i, P \right) > R_s \end{cases}$$

Here $p\left(s_i, P \right)$ is the probability of sensing the object at point P by s_i.

Particularly, the model being a Boolean model is known as *binary disc sensing model*.

However, in order to provide a more realistic model to deal with the uncertainty in the sensing detection, the aforementioned model was rectified in Zhu et al. (2012) as the *probabilistic disc sensing model* as follows:

$$p(s_i, P) = \begin{cases} 1, & d(s_i, P) \le R_s - R_e \\ e^{-\omega \alpha^\beta}, & R_s - R_e < d(s_i, P) \le R_s + R_e \\ 0, & d(s_i, P) > R_s + R_e \end{cases}$$

where $\alpha = d(s_i, P) - (R_s - R_e)$ and ω and β are parameters affecting the detection probability when the object located at point of interest (PoI) lies between $(R_s - R_e)$ and $(R_s + R_e)$ with R_e ($< R_s$) being the uncertainty measure in sensor detection, as explained in Figure 14.1.

The models explained in Figure 14.1 are collectively known as the *uniform sensing model*.

Irregular Sensing Model

The disc sensing model is too ideal and does not consider the environmental factors, viz. obstacles etc., while formulating the sensing problem. Therefore, a more realistic sensing model known as *irregular sensing model* has come into existence which considers all such factors, such as environmental obstacles, weather conditions, etc., affecting the overall sensing process (Figure 14.2).

Mobility Models

With regard to the mobility, there comes two important notions: mobility pattern and mobility model. Mobility pattern deals with how people and things move, whereas mobility model deals with the mathematical generalization of the characteristics of mobility patterns. A mobility model defines how to compute location,

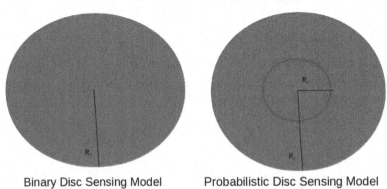

Binary Disc Sensing Model   Probabilistic Disc Sensing Model

Figure 14.1   Disc sensing model.

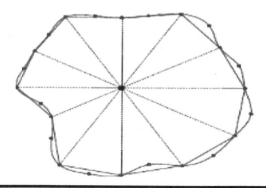

Figure 14.2   Irregular sensing model. (From Mahamed et al., 2017.)

speed, and acceleration of the sensor nodes in the network over time. As already explained in Chapter 12, mobility models can be broadly categorized as trace-based and synthetic models. *Trace-based* mobility models refer to the mobility patterns of real-life objects which are kept under constant observation over a long period of time. Though traces provide accurate information about the movement pattern of the objects, capturing the movement trajectory of mobile nodes is quite difficult even when adequate amount of previous historical data is available from the recurrent mobility patterns. Now, here comes the synthetic model into picture. *Synthetic models* aim for realistic mapping of the movement pattern of the mobile nodes without any use of traces (Camp et al., 2002).

Synthetic models are further classified into two subcategories: entity and group mobility models, each having further classification as in Figure 14.3.

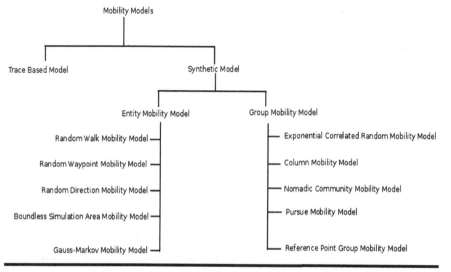

Figure 14.3   Classification of synthetic mobility models.

Entity mobility models, also known as individual mobility models, describe how nodes in the network can move independent of one another, i.e. each node can move without any consultation of others, whereas group mobility models describe the scenario where nodes' movements are dependent on one another, as in military applications, where nodes collaborate with each other in order to achieve some well-specified common goal.

Moreover, each of the variants of individual and group mobility models have already been discussed in greater depth in the aforementioned Chapter 12.

Coverage Taxonomy

However, the Nomenclature section has already discussed a brief classification of coverage above; a more detailed classification is provided in Figure 14.4.

As shown in the Figure 14.4, coverage is broadly classified either according to its degree or on the basis of area covered.

Based on the number of nodes employed in the field of interest, coverage is said to be 1-coverage or k-coverage. If the entire FoI is covered by just one node, it is said to 1-coverage whereas if the FoI contains k number of nodes to cover its every point, it is said to have a k-coverage.

Moreover, based on how much area is being covered in the FoI (i.e., Θ-coverage), coverage can be classified as full coverage and partial coverage. Full coverage is defined as each point in the FoI is under the observation of at least one node; partial coverage requires only a particular portion of the FoI to be covered. Furthermore, partial

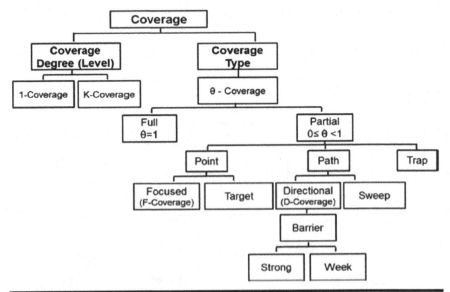

Figure 14.4 Coverage taxonomy in sensor networks. (From Mohamed et al., 2017.)

coverage can be of three types: point, path, and trap coverage. Here, point coverage requires only a particular point to be covered whereas path coverage refers to covering a definite path to be covered; trap coverage guarantees any moving object to be detected before it moves a priori known displacement. Point coverage is further divided into focused and target coverage whereas path is divided into directional and sweep coverage. Each of the aforementioned types is discussed in more detail as follows.

Θ-*Coverage*

As indicated above, the Θ-coverage requires a certain percentage of the field of interest to covered by the sensor nodes deployed. Θ can be found between [0, 1]. If Θ ≥ 0, but < 1, it is said to be partially covered, and when Θ = 1, coverage is said to be full coverage.

Full Coverage

As stated earlier, the full coverage also known as blanket coverage requires each point in the field of interest (FoI) to be brought under observation using 1 or more number of nodes. In the other way, full coverage can be defined as the one in which every point is covered by using at least one node.

More illustratively, if the network contains n number of nodes, say $s_1, s_2, s_3, \ldots, s_n$ and S be the set of these nodes such that $S = (s_1, s_2, s_3 \ldots, s_n$ such that $s_i \in S\}$. Let A_R be the total area of the sensing field and s_A be the region covered by a sensor s. Then the full coverage is said to be achieved if

$$\coprod_{s \in S} s_A \geq A_R$$

Similarly, if a point, say p, is being covered by one or more nodes in the network and $C_s(p)$ refers to this phenomenon, then the total coverage of this point p denoted by $C(p)$ can be defined as follows

$$C(p) = \sum_{i=1}^{k} C_s(p)$$

And a total coverage of degree k is said to be obtained if for every point p in the FoI,

$$C(p) = k$$

Partial Coverage

Partial coverage is a variant of Θ-coverage which does not require the entire sensing area to be covered but only a definite portion. Applications viz. environmental monitoring requires coverage only to a certain degree. More illustratively, a deployment or sensing area or field of interest is said to be Θ-covered if,

$$A_R \cap \left(\coprod_{s \in S} s_A \right) \geq \Theta(A)$$

where Θ(A) is region to be covered as per the application requirement.

Point Coverage

As indicated above, point coverage requires only a particular point to be covered by a node. Moreover, point coverage has further two variants: focused coverage and target coverage.

Focused Coverage Focused coverage requires the region close to the PoI to monitored with the priority higher than that for the region distant from PoI.

Target Coverage Target coverage pays all its attention to a particular object of interest.

Path Coverage

Path coverage requires a set of points to monitored which are aligned along a path crossing the width of the area of deployment or FoI (Kumar et al., 2005). A path coverage can further be categorized either as directional coverage (d-coverage) or sweep coverage.

Directional Coverage Directional coverage aims at detecting the event of intrusion between the two boundaries through which the trespasser attempts to penetrate. It detects the penetration through the first boundary before the second one is penetrated (Bai et al., 2009).

The next variant of directional coverage is the *barrier coverage* which detects the event of intrusion irrespective of their penetration direction. A barrier coverage can be either strong or weak. A strong barrier coverage ensures the detection of intruders whatever path they choose, whereas weak barrier coverage determines the trespassing when the object passes along congruent paths (Kumar et al., 2005).

Sweep Coverage Sweep coverage is another variant of path coverage in which a set of points (of interest) are monitored periodically instead of continuously (Li et al., 2011).

Trap Coverage An unsupervised deployment (random deployment) of the network may result in a number of coverage holes, i.e., the uncovered regions in the

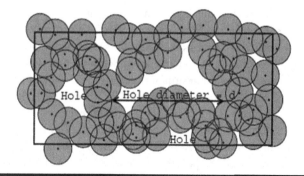

Figure 14.5   d-Trap coverage. (From Balister et al., 2009.)

network. Let *d* be the largest diameter of coverage hole as depicted in Figure 14.5, then *d-trap coverage* in the network ensures every moving object or event of interest is detected (or covered) for every displacement of d with definite certainty.

Coverage Deployment Strategy

Due to its direct relation with the resource optimization in the network, coverage has always been a prime issue of concern since the evolution of sensor networks. When dealt generously, maximization of coverage while minimizing the number of nodes may result in cost-effective and energy-efficient network deployment. Based on the presence of mobility in the network, coverage deployment strategy can be divided into two categories: static and dynamic. As implied, static coverage deployment strategy involves only unmovable static nodes, whereas dynamic involves mobile nodes. The problem formulation for the static coverage can be easily understood with the popular art gallery problem (O'Rourke, 1987) which aims at finding the minimum number of observers to be placed in an art gallery room so that every point in the room is seen by at least one observer.

Moreover, since the emphasis in this chapter is on sensor networks featured with mobility, dynamic coverage is discussed in more detail as follows.

Dynamic Coverage

Contrary to the static coverage, dynamic coverage includes some or all the nodes with the ability to change their locations from one place to another and, hence, are more complex to handle when compared with their static counterpart. However, mobile nodes are often needed in order to provide better monitoring and surveillance, especially in the applications with random deployment of nodes. For example, when the nodes are dropped by an aircraft or projected through some other means in the human-inaccessible and hostile network environment, nodes' deployment can never be planned. Such applications may suffer with a lot of coverage holes, i.e., uncovered regions. Mobile nodes can play a very significant role in these applications as they may be directed to relocate themselves to improve the overall coverage.

In the networks with mobile nodes, a network region or a point in the network may alternate between covered and uncovered status, i.e., when visited by a roaming node, it can be reported; otherwise, it remains unreported. Mobility in sensor networks has already been explored widely in the context of coverage.

In Liu et al. (2005), the dynamic aspect of coverage has been studied well by inspecting the random mobility model thoroughly for the applications in which simultaneous coverage of all the points are not required at a specific time instance. It is observed that when coverage is defined as the covered area during a particular time interval, it is improved, i.e., when the instantaneous mean coverage is considered to be unchanged, a larger area can be covered by employing mobility among the nodes in network.

Similarly, when explored in the context of heterogeneous wireless sensor networks, dynamic coverage, modeled as a control problem towards the computation of best paths while preserving the connectivity, responds well by achieving the intended objective (Gabriele and Giamberardino, 2008).

Mobility increases the design complexity and proposes a new set of challenges; it has been identified as a tool for achieving improved coverage and connectivity in a wide variety the sensor networks' applications.

References

X. Bai, L. Ding, J. Teng, S. Chellappan, C. Xu and D. Xuan, Directed coverage in wireless sensor networks: Concept and quality. In: *IEEE 6th International Conference on Mobile Adhoc and Sensor Systems*, 2009, pp. 476–485.

P. Balister, Z. Zheng, S. Kumar and P. Sinha, Trap coverage: Allowing coverage holes of bounded diameter in wireless sensor networks. In: *IEEE INFOCOM*, 2009, pp. 136–144.

T. Camp, J. Boleng and V. Davies, A survey on mobility models for ad hoc network research. *Wireless Communications and Mobile Computing*, vol. 2, pp. 483–502, 2002.

S. Gabriele and P. D. Giamberardino, The area coverage problem for dynamic sensor networks. *New Developments in Robotics Automation and Control*, 2008.

S. Kumar, T. H. Lai and A. Arora, Barrier coverage with wireless sensors. In: *The 11th Annual International Conference on Mobile Computing and Networking*, 2005, pp. 284–298.

M. Li, W. Cheng, K. Liu, Y. He, X. Li and X. Liao, Sweep coverage with mobile sensors. *IEEE Transactions on Mobile Computing*, vol. 10, no. 11, 2011, pp. 1534–1545,.

X. Li, H. Frey, N. Santoro and I. Stojmenovic, Focused-coverage by mobile sensor networks. In: *IEEE 6th International Conference on Mobile Adhoc and Sensor Systems*, pp. 466–475, 2009.

B. Liu, P. Brass, O. Dousse, P. Nain and D. Towsley, Mobility improves coverage of sensor networks. In: *Proceedings of the 6th ACM International Symposium on Mobile Ad Hoc Networking and Computing*, 2005, pp. 300–308.

Y. Liu and W. Liang, Approximate coverage in wireless sensor networks. In: *The IEEE Conference on Local Computer Networks (LCN)*, 2005, pp. 68–75.

Y. Mei, C. Xian, S. Das, Y. C. Hu and Y.-H. Lu, Sensor replacement using mobile robots. *Computer Communications*, vol. 30, no. 13, pp. 2615–2626, 2007.

S. M. Mohamed, H. S. Hamza and I. A. Saroit, Coverage in mobile sensor networks (M-WSN): A survey. *Computer Communication*, vol. 110, pp. 133–150, 2017.

J. O'Rourke. *Art Gallery Theorems and Algorithms*. London, UK: Oxford University Press, 1987.

C. Zhu, C. Zheng, L. Shu and G. Han, A survey on coverage and connectivity issues in wireless sensor networks. *Journal of Network and Computer Applications*, vol. 35, no. 2, pp. 619–632, 2012.

FAULT-TOLERANT WIRELESS SENSOR NETWORKS

Chapter 15

Fault Tolerance in Wireless Sensor Networks

Since their inception, wireless sensor networks (WSNs) have emerged as an obvious choice for a variety of applications viz. military operations, agricultural monitoring, patient monitoring, structural monitoring, intruder detection, habitat monitoring, etc. In such applications, WSNs are deployed for information collection, event detection, or target tracking over a well-defined zone of interest. And these objectives are achieved by sensing and transmitting the expected information to the processing centers for further processing. These applications have already been benefitted a lot by the unmatched capability of WSNs, but being a variant of challenged networks, WSNs suffer with a number of problems, e.g., limited life expectancy, limited resources, asymmetric data rate, etc. Limited life expectancy mainly refers to the limited power with which the nodes are equipped. In most of its major applications, sensor nodes' deployment restrict the replacement of their battery upon depletion as the sensing area or the zone of interest might be human inaccessible. Along with these constraints, the hostile and harsh deployment environment (e.g., rain, thunder, cyclone, snow, etc.) further complicate the scenario by causing frequent and unexpected errors in the system; such errors may result in the failure of the nodes or sometimes the failure of the network too. Therefore, these problems necessitate the provision of solutions to mitigate the effect of these inadequacies to empower the WSNs to achieve their intended objectives even in the presence of faults. In the literature, the ability of fulfilling the expected functionality in the faults' presence is known as fault tolerance, and the concerned networks equipped with this ability are called fault-tolerant wireless sensor networks.

Provisioning fault tolerance in the network requires a complete workout from fault detection to fault recovery. Fault detection states the need of identifying the faults which could lead to the system failure; fault recovery refers to the way out and bringing the system back into action to pursue the intended functionality. Obviously, a fault can never be detected if one is not familiar with the possible sources of faults; hence, this chapter first discusses the various possible sources of faults in the wireless sensor networks leading to system error or failure. Then, in the subsequent sections, fault detection and fault tolerance will be detailed with a few of their respective manifesting schemes in the literature.

Classification of Faults and Failures in WSNs

Before getting into the depth of fault detection and fault recovery (i.e., into the fault tolerance) one is required to be familiar with the terminology being used in the literature. Therefore, this terminology is explained below for better understanding of the subject.

Fault: In the context of the wireless sensor network, any condition which causes or leads the system to an erroneous state is termed to be a fault.

Error: Error refers to the system state which is undefined and incorrect and may lead to the system failure.

Failure: Failure manifests the error. It represents the system inability or the system deviation from its intended objective, i.e., system/component has failed to execute the claimed function as per its specification.

Fault Tolerance: Fault tolerance refers to the ability of the system to pursue its intended objective even in the presence of faults/errors. It ensures that system/component perform the expected functionality despite being affected by faults/errors present in the system.

Fault Detection: Fault detection consists of revealing the presence of faults in the system.

Fault Recovery: Fault recovery refers to the revival of the system's expected functionality after detecting the fault by rectifying the failed unit. Moreover, redundancy and replication have been implemented very popularly as the fault recovery methods in the sensor networks (Liu et al., 2009).

The above cited terms can be explained in a more detail with the help of an exemplary scenario, as follows:

Let us think of a network in which to meet the application requirements; deployed nodes are equipped with multiple sensors to measure the different environment parameters viz. temperature, humidity, noise, etc. The nodes are programmed to forward their sensed data to a designated node, say P, in the network for further processing so that the processed data can be sent to the sink/base station for the next level of processing or for some decision-making process as required by the application. Let the harsh network surrounding cause a node, say Q, have an impact, loosening the connection of the node to one of its sensors. Since the node Q is not provisioned to handle such situation, an erroneous state arises when the node tries to retrieve the data from its loosened sensor to forward the same for further processing at the specially designated node P. The sensed data fails to arrive at the node P within the specified duration, resulting in the failure of the node Q in view of P.

Here, loosening the connection to a sensor due to an impact is termed as a *fault* and attempting to retrieve data from the disconnected sensor brings an *error*. Failing in sending the environment data to the designated node in a specified time interval results in the *failure*; detecting the faulty functionality of the node is termed as *fault detection*. Reviving the correct functionality by invoking some redundant node, by replicating the system component required for the intended operation, or by any other mean is termed as *fault recovery*. The composite effect of fault detection and fault recovery makes the network resilient to the faults even though they are there in the network. Hence, the network becomes a *fault-tolerant* network, improving the network reliability.

Classification of Sources of Faults

Wireless sensor networks comprise a number of hardware and software components and the application for which the WSN is deployed. Each of these components is subjected to become faulty for numerous reasons. Also, the surrounding in which network is deployed contributes a lot to the faults which the network has to suffer with. The following layered classification attempts to cover almost every possible source of faults so that preventive or curative actions could be taken developing a fault-resilient network (Akyildiz et al., 2002).

In the layered classification of the sources of faults in WSN, it is to be categorized into three major layers- Hardware, Software, and Application layer (Figure 15.1).

Hardware Layer

As stated earlier, every network component is subjected to some inherent or unavoidable defects, leading to errors and failures in turn; there can be three basic reasons for this: low-quality components, power restriction, and hostile environment. Low-quality components may bring the network to a down state as soon as it starts to operate. Power restriction refers to the problem when nodes might fail to sense the surrounding accurately when energy dips below a certain level or the nodes might even fail to operate. The hostility of the environment in which the network has been deployed may also make the network suffer as the thunder, storms, lightning, floods, etc. might severely affect the network performance.

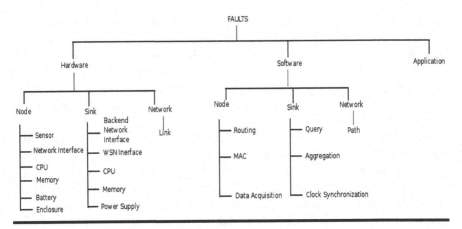

Figure 15.1   Classification of faults' sources in WSN.

This layer covers each of the faults which is caused by the malfunctioning of any of the hardware components; moreover, such hardware faults can further be categorized as node faults, sink faults, and the network faults, depending on the nature of malfunction.

Node Faults

Among the three major components of any sensor network—node, sink, and the network—node is the most significant one constituting the wireless sensor networks. Under the hardware layer of the faults, node faults are the faults which are caused by the malfunction of the unit a node consists of. A sensor node comprises a number of units viz. sensors, CPU, memory, network interface, battery, the enclosure, etc. Dereliction of any of such components may contribute to the failure of the node. More illustratively, the hostile environment of the network deployment may cause the enclosure of node to suffer with an impact, exposing the node's hardware to extreme environmental conditions, i.e., *enclosure faults*. Similarly, *the battery faults, CPU faults, memory faults, network interfacing fault*, and *sensors faults*.

Sink Faults

Sink is also an important component of WSN which is responsible for the delivery of data collected from the deployed nodes representing different environment parameters of the concerned sensing area. The significance of this component can be easily seen from the fact that failure of this single entity lets the entire network down, as the data from the sensing nodes can never reach the back-end system to participate in any sort of decision-making process.

Like sensing nodes, sink is also comprised of a number of hardware components subjected to become faulty viz. CPU, memory, power unit, WSN interface, and the back-end network interface. Malfunction of any such node may lead to the failure of the sink, i.e., the *sink fault*. More illustratively, network interface may not be available to the fields where sink is to be deployed; hence, satellite connection may be provided as an alternative solution to this. However, hostile environment, for instance, severe thunderstorms or lightning might result in unavailability of connection. Similarly, the interruption in the power supply may turn the sink dead and hence result in the failure of the network.

Network Faults

The collected data is to be transported to base station, destined to the end user for some application-specific decision-making process; this transportation is done via the communication network. This is the network which performs routing in order to make data available to the sink. Communication links are the essential parts of the network which carries all the data, and the links are highly volatile in nature. There may be numerous reasons that may cause a link to malfunction, e.g., mobility of the nodes, obstructions in radio interface, and hence the *network fault*. More illustratively, nowadays, most of the sensor applications are incorporating the moving nodes, and the change in the location of nodes causes instability of the links; likewise, radio interfaces of the nodes may get blocked due to the unexpected and unpredictable natural scenario, and such obstructions in the link establishment may lead the network fault.

Moreover, there is an additional aspect of classifying the hardware faults in WSN which is based on the duration of faults, i.e., faults' temporal aspects. As per this classification, there are four classes of faults: transient, intermittent, permanent, and potential.

Faults induced by the surrounding which are less intense and are of much shorter duration are called *transient faults*. *Intermittent faults* are the ones which are repetitive in nature, i.e., they appear and disappear; such faults are majorly caused by the non-environmental conditions, viz. loosening of sensors' connections, etc. *Permanent faults*, as implied by the name, refers to the faults caused by non-functioning of hardware components, and they can be only be cured by fixing or substituting the faulty component. Moreover, faults due to the diminishing resources are called *potential faults*.

Software Layer

This layer mainly covers the faults which are either caused by the system software viz. operating system or by the middleware, including routing, data acquisition, aggregation, etc. The prime sources of faults under this category are the software bugs. Moreover, a hardware fault/failure may also lead to the software fault/failure, for example, as soon as the node's energy starts getting below a certain level, sensors attached to the node may generate inaccurate data and hence the error.

Bugs can be found in every components of the sensor network viz. in the nodes, in the network, and in the sink. Bugs present in the nodes may lead to inefficient routing, inappropriate medium access, inaccurate data acquisition, and improper aggregation, especially when clustering approach is implemented for routing.

When a communication network suffers with the problem of software bugs, the network may result in incorrect routing of the message, i.e., the *path error.*

Since sink is the central repository of the data for the access to end user, software bugs may result in the loss of data within the period of fault occurrence, which is definitely an irreparable loss.

Application Layer

Wireless sensor network is an application-specific network, and, obviously, different applications have their own specific requirements. This layer covers every fault which is caused by either some external activities or due to some internal activities that may lead to the failure of the entire system.

Classification of Failures

As explained above, the faults lead to failure where the affected system deviates from its expected duty. For example, the mobility of nodes, if not handled adequately, may result in link failure; diminishing of battery installed into the nodes may cause inaccurate reading, hence the failure; also, an unauthorized access may turn the sink to receive irrelevant readings, causing it to produce incorrect results. In the available literature, failures have been classified as follows (de Souza et al., 2007) (Figure 15.2).

Omission Failure

When a service responds intermittently only and becomes out of reach very frequently, it is said to suffer with omission failure. Such failures are generally caused by network transmission error or inadequate intermediary buffer spaces to hold the message being transmitted.

Figure 15.2   Failures' classification in WSN.

Crash Failure

Crash failure refers to the permanent stoppage of the service, i.e., at some point in its operation, the service stops responding to the service requests.

Timing Failure

Wireless sensor networks are the most vital example of application-specific networks in which, along with the correctness of values, delivery timing is also of great significance. This category of failures represents behavior of the systems in which services fail due to delayed or early delivery of the messages concerned. As the data arrives out of time, application might fail to process them; hence the failure.

Value Failure

When a service fails due to a set of incorrect measurements, i.e., incorrect data, the system is said to suffer with value failure. More illustratively, although the data has arrived timely, the inaccuracy of the data may lead to an unexpected behavior of the system; hence the failure.

Arbitrary Failure

All such faults which cannot be categorized in any of the above cited classes belong to this class of arbitrary failure. For example, when a query could not be forwarded to the concerned node despite the sender being acknowledged; in the aggregation process, at some cluster head (in some clustering routing protocol), both the incorrect and the correct one have been processed, etc.

Fault-Detection Techniques

Fault detection implies determining faulty functionality of the components of a system, i.e., to determine whether the service being offered is as per the specification of the system. Also, it may be used to declare whether the services would continue in the near future or not. In the standard literature, there are numerous fault detection techniques meeting the application-specific requirements of the wireless sensor networks which can be classified in two major aspects: who detects the fault and how the faults are detected. Both aspects are discussed in the following sections.

Involvement-Based Fault Detection

Depending on who is participating in the fault detection, classification can be done as follows (Figure 15.3).

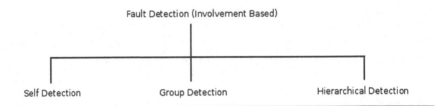

Figure 15.3 Fault-detection techniques based on involvement of nodes.

Self-Detection

In some applications, nodes in the network are able to detect the presence of faults, as they are equipped with devices/units capable of doing so. For example, in the networks deployed in some area that is heavily affected, for instance, by frequent thunderstorms, accelerometers may get installed into the nodes to determine whether the impact to the node is sufficient to cause some hardware malfunction (Harte et al., 2005); the change in the location of a node can be easily be determined by keeping track of the neighbors' identity; by installing devices measuring the current voltage level, nodes can determine if the battery is about to be exhausted (Benini et al., 2000; Rakhmatov and Vrudhula, 2001).

Group Detection

Group detection refers to the class of techniques in which detection is achieved as a result of collaborative efforts by a number of nodes in the network, i.e., the behavior of the node probable to become faulty is monitored by a group of nodes; hence the name. For example, identifying a node generating inaccurate values can be done if it is to be compared with a reference value, as the nodes falling into the same region and under close proximity are supposed to generate similar values (Ding et al., 2005; Krishnamachari and Iyengar, 2004).

Hierarchical Detection

Hierarchical detection represents the class of techniques in which a hierarchy of the nodes is maintained to identify the presence of faulty nodes. Also, it has become a very common practice to appoint sink at the top of the hierarchy to identify the faulty nodes. For example, in his proposal of a health monitoring system—Memento—S. Rost et al. devised a network topology to propagate the detection result from the child node to its parent node and up to the base station, i.e., each node monitors its immediate children and forwards the result to its parent node to be propagated ultimately to the sink (Rost and Balakrishnan, 2006); in a network, every node learns the network topology of its vicinity and forwards the same to the base station, which prepares the complete picture, i.e., the complete network topology to be used in sending the route updates if it finds some failed nodes (Staddon et al., 2002).

Approach-Based Fault Detection

Another classification of fault detection techniques in the literature is based on how the faults are detected, i.e., approach-based classification (Muhammed and Shaikh, 2017). The classification is depicted in Figure 15.4 and is elaborated as follows:

In a broader way, this classification comprises three main categories: centralized, distributed, and hybrid approaches. As implied by the name, the centralized approach provisions a single node, preferably the base station, to identify and weed out the faulty functional unit from the network. On the contrary, the distributed approach makes each of the nodes in the network monitor and analyze the status of nodes. Such algorithms are run on every node to produce the local status, which is propagated through the network. The hybrid approach combines the concepts of both the aforementioned classes in view of giving a better solution.

Centralized Approach

As stated above, in the centralized approach, the specially designated node or the sink takes care of analyzing the status of remaining nodes in search of a faulty node. Centralized approaches can be further classified in two subcategories: statistics-based approach and soft computing centralized approach.

In the statistics-based approach, statistical techniques such as sigma test and mean are used by the analyzing node to figure out the faulty node in the network. However, in the soft-computing-based approach, machine learning techniques are used by the monitoring node which can further be categorized into supervised and unsupervised learning. In supervised learning techniques, (already existing) training data sets are used to learn the difference between the expected and unexpected data, i.e., between the actual and faulty data; however, in the unsupervised learning, no such training data sets are used.

Distributed Approach

Distributed approach requires the detection algorithms to be run on every node in the network. Success of the distributed approach lies in the fact that the data from wireless sensor networks have strong spatial and temporal correlation, i.e., any instantaneous data from the network has a strong relation with its previous and succeeding data, and, also, the data obtained from the nodes in close proximity has a higher degree of similarity. Moreover, the distributed approach can be further categorized into six sub-categories: neighborhood-based, statistics-based, probability-based, soft-computing-based, self-detection, and cloud-based detection.

> *Neighborhood-based*: Neighborhood-based detection is a distributed approach that makes full use of the spatiotemporally correlated nature of sensor data. The detection is done in any of the two ways, *majority voting* or *weighted*

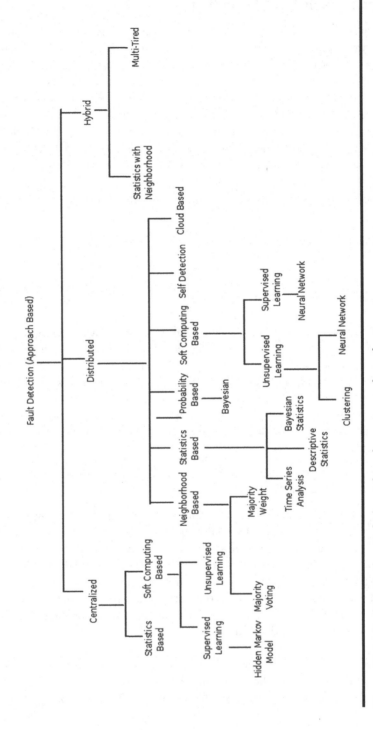

Figure 15.4   Fault-detection techniques based on approach made.

majority voting (Chen et al., 2006). In the majority voting technique, the faultiness of a node is determined by the majority of fault status received by the neighbor nodes, whereas in the weighted majority voting method, each network node is assigned a particular weight, which is used to compute the weighted majority to confirm a node to be faulty.

Statistics-based: In the statistics-based techniques, statistical methods are used to pursue the search of faulty nodes. The scheme can further be classified into three subclasses: time series analysis, descriptive statistics, and Bayesian statistics method.

Time series analysis utilizes statistical test viz. Kolmogorov–Smirnov test and Kuiper test on the time series data to weed out the outliers in determining the faulty nodes.

Descriptive statistics uses pure statistical method such as applying central tendency measures—mean and the median of the neighbor nodes—to identify the faults. Lastly, in the Bayesian statistical method, Bayes' theorem is applied to compute the likelihood of a node becoming faulty.

Probability-based: In the probability-based technique, the posterior fault probability of a node is computed to predict the final status of nodes in terms of failure probabilities of the node and their respective neighborhood nodes.

Soft-computing-based: The soft-computing-based techniques consist of utilizing the machine learning techniques, which can be further divided into two classes (as also under the centralized approach): supervised and unsupervised learning. In the supervised learning technique, training data sets are used to differentiate between actual and faulty data to determine the presence of faults. Neural-network-based techniques are used to implement the above, i.e., to predict the faultiness in data. However, in the unsupervised learning technique, no training data sets are used, but the approach is quite similar to the statistical problem of density estimation. Basically, the clustering technique of dividing the nodes into an optimal number of clusters where each cluster is accompanied with a cluster head to perform the fault-analysis task and certain neural networks viz. self-organizing maps (SOMs) are used to implement the idea of unsupervised learning.

Self-detection: The self-detection technique utilizes the spatiotemporally correlated nature of the sensor data as it takes help from the nodes in its close vicinity in determining its faulty functionality. However, the final decision depends on the node itself in finding its faulty status.

Cloud-based detection: As implied by name, cloud-based detection techniques employs the cloud infrastructure in order to reduce computational complexity. The sensory data is deported to the cloud infrastructure where schemes like map-reduce is used in achieving the objective of fault detection. This not only lessens the use of local and resource-constrained network devices but also reduces fault-detection time.

Hybrid Approach

As stated earlier, hybrid algorithms combine the efforts from both aforementioned categories: centralized and distributed. Schemes belonging to the hybrid class of fault detection techniques finds their suitability to the multi-tiered network architectures viz. clustering based sensor networks or they refer to a combination of various detection schemes listed above. Hybrid algorithms are best suited to the cluster-based network architecture where nodes are divided into a number of clusters, each being represented by a well-designated node (generally rotating among the nodes) called the cluster head. The cluster head collects the data from its cluster and forwards the same to the base station for analysis; here, it is distributed, because each node is participating, and centralized, because analysis is being performed at some central node in the network, say sink, and in this way, the hybrid one. Similarly, when multiple schemes are combined together, the approach becomes the hybrid one.

Classification of Fault-Tolerance Techniques

Fault tolerance refers to an important property of the system where system is capable of preserving its functioning even in the presence of faults. Faults do occur, but their effects are alleviated through some provisioning such that the system is not required to compromise its intended objective. In other words, the system does deviate from its intended services in the presence of faults. The fault tolerance techniques available in the literature can be majorly classified into two ways: temporal classification and objective-based classification (Chouikhi et al., 2015).

Temporal Classification

Depending upon when the fault-tolerance schemes are being activated/triggered, temporal classification comes into shape, i.e., the classification is made on the basis of whether the scheme is being activated a priori or after the occurrence of fault. When fault-tolerance procedures are provisioned a priori, i.e., faults are yet to be detected, techniques are categorized as *preventive techniques*, and when they are made to run after the occurrence of fault, schemes are categorized as *curative techniques*.

Preventive Fault-Tolerant Techniques

As implied by the name, the schemes making wise use of available resources to mitigate the possibility of occurrences of faults are classified as preventive techniques. More illustratively, fault-tolerant techniques utilizing the network resources efficiently, along with using all the other alternatives providing the same services uninterruptedly, can be classified as preventive fault-tolerant techniques. Furthermore, preventive techniques can be found being implemented at two different levels: node level and network level; however, at both the levels, maximizing the network lifetime is attempted so that the desired services can be continued uninterruptedly via applying sleeping strategy, creating effective task scheduling, provisioning even load distribution, etc. At node level, efforts are made to prolong the node lifetime, ultimately contributing to the network lifetime, and at network level, the failures of a few nodes may be tolerated without interrupting the services.

Curative Fault-Tolerant Techniques

On contrary to the preventive techniques, in curative fault-tolerant techniques, procedures take into account the occurrence of failures, i.e., the schemes are activated when the occurrence of some failure is confirmed by the fault-detection techniques. Once the faults are reported, efforts are made to resume the sensing or transmission by healing the failed components. Here, failed nodes are not always replaced, but, instead some redundant nodes are activated to take care of the task allocated to the failed nodes so that service will not be affected.

Objective-Based Classification

Objective-based classification refers to techniques which focus on some well specified objective, like energy. There are four main classes under this and the same is depicted in Figure 15.5 as follows:

- Energy-centric
- Flow-centric
- Data-centric
- Quality-of-Service (QoS)-centric

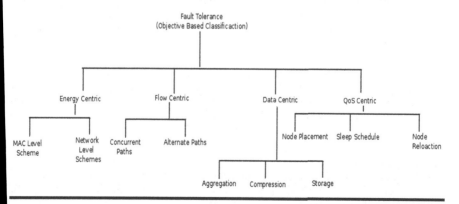

Figure 15.5   Objective-based classification of fault-tolerance schemes.

Energy-centric fault-tolerant techniques: Such schemes achieve fault tolerance by overcoming the problem of premature battery depletion in the sensor nodes by applying various energy-saving procedures at different layers, for example, the MAC layer, network layer, etc.

Flow-centric fault-tolerant techniques: Flow-centric schemes deal with establishing robust linkages between the nodes and the sink/base station to ensure smooth flow of data between the participating nodes. Moreover, to alleviate the effect of path failures, alternate paths are used.

Data-centric fault-tolerant techniques: Since data is the most valuable resource satisfying the need of concerned application of wireless sensor networks, it has to be well taken care of. These procedures focus on data-manipulating schemes to reduce the total network data-flow without affecting the quality of data. Moreover, the since data being communicated is minimized, communication cost is also saved, hence contributing to prolonging network lifetime.

QoS-centric fault-tolerant techniques: Schemes belonging to this class try to attain fault tolerance while ensuring some important Quality-of-Service (QoS) parameters such as coverage and connectivity, where coverage determines how well a point of interest in the sensing field is being covered and connectivity refers to how well the data is being reported to the data collector.

References

I. F. Akyildiz, W. Su, Y. Sankarasubramaniam and E. Cayirci, A survey on sensor networks. *IEEE Communications Magazine*, vol. 40, pp. 102–114, 2002.

L. Benini, G. Castelli, A. Macii, E. Macii, M. Poncino and R. Scarsi, A discrete-time battery model for high-level power estimation. In: *Proceedings of the Design, Automation and Test in Europe Conference and Exhibition*, 2000, pp. 35–39.

J. Chen, S. Kher and A. Somani, Distributed fault detection of wireless sensor networks. In: *Proceedings of the 2006 Workshop on Dependability Issues in Wireless Ad Hoc Networks and Sensor Networks, DIWANS '06*, pp. 65–72.

S. Chouikhi, I. El Korbi, Y. Ghamri-Doudane and L. A. Saidana, A survey on fault tolerance in small and large scale wireless sensor networks. *Computer Communications*, vol. 69, pp. 22–37, 2015.

L. M. S. de Souza, H. Vogt and M. Beigl, A survey on fault tolerance in wireless sensor networks. *Internal Report*. Faculty of Computer Science, University of Karlsruhe, Germany, 2007.

M. Ding, D. Chen, K. Xing and X. Cheng, Localized fault-tolerant event boundary detection in sensor networks. In: *INFOCOM*, 2005.

S. Harte, A. M. Rahman and K. M. Razeeb, Fault tolerance in sensor networks using self-diagnosing sensor nodes. In: *The IEEE International Workshop on Intelligent Environment*, 2005, pp. 7–12.

B. Krishnamachari and S. Iyengar, Distributed Bayesian algorithms for fault-tolerant event region detection in wireless sensor networks. *IEEE Transactions on Computers*, vol. 53, 2004, pp. 241–250.

H. Liu, A. Nayak and I. Stojmenovic, Fault tolerant algorithms/protocols in wireless sensor networks. In: *Handbook of Wireless Ad Hoc and Sensor Networks*, 2009, pp. 261–291.

T. Muhammed and R. A. Shaikh, An analysis of fault detection strategies in wireless sensor networks. *Journal of Network and Computer Applications*, vol. 78, pp. 267–287, 2017.

D. Rakhmatov and S. B. Vrudhula, Time-to-failure estimation for batteries in portable electronic systems. In: *Proceedings of the 2001 International Symposium on Low Power Electronics and Design*, 2001, pp. 88–91.

S. Rost and H. Balakrishnan, Memento: A health monitoring system for wireless sensor networks. In: *SECON*, 2006.

J. Staddon, D. Balfanz and G. Durfee, Efficient tracing of failed nodes in sensor networks. In: *Proceedings of the 1st ACM International Workshop on Wireless Sensor Networks and Applications*, 2002, pp. 122–130.

CROSS-LAYER
OPTIMIZATION

Chapter 16

Cross-Layer Optimizations in Wireless Sensor Networks

With the ongoing advancements in radio technology and that of in micro-electro-mechanical technologies, it has been possible to devise tiny-size battery-powered devices with low power consumption which can sense their surroundings and communicate to a centralized base station or sink after performing a little processing over the sensed data for further course of application-oriented actions as depicted in Figure 16.1. An end user can retrieve data from the sink using wireless connectivity.

Wireless sensor networks (WSNs) have evolved as a very popular platform for the development and deployment of a wide variety of applications to benefit humanity viz. military operations, agricultural monitoring, patient monitoring, structural monitoring, intruder detection, habitat monitoring, etc. In all these aforementioned applications, WSN is used for information collection, event detection, or target tracking in a definite and well-defined zone of interest, and with its unmatched capability, WSN has benefitted these applications a lot.

Wireless sensor networks are deployed into different environments depending on the very nature of applications, i.e., on the basis of purpose, a WSN can be deployed into a completely different network scenario. For example, in order to capture the audio/video information from a field, sensors capable of sensing multimedia information are deployed and the network is called wireless multimedia sensor networks (WMSNs); when sensors are deployed inside an industry for study and monitoring purposes, the network is called an industrial wireless sensor network (IWSN); when sensors are installed on patients' bodies for healthcare and regular

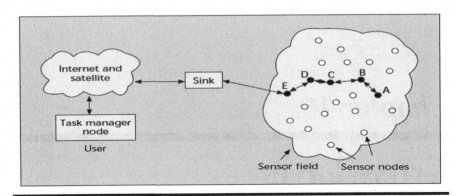

Figure 16.1   Wireless sensor network. (From Akyildiz et al., 2002.)

monitoring purposes, the network is called wireless body area network (WBAN); when sensors are instructed to take a particular course of action based on a match with the preconfigured conditions, the network is called wireless sensor and actuator network (WSAN); similarly, when nodes are equipped with mobility features, it is called a mobile sensor network (MSN), and if nodes are able to move with high speed, the network is particularly known as vehicular sensor network (VSN).

With such a large variety of sensor networks and vast number of applications, numerous challenges have been presented before the researchers' community and, thus, a number of different solutions have come into existence suited to different problems.

However, some problems may be found in every instance of sensor networks, e.g., requirement of energy-efficient solutions. Though the advances in radio technology and micro-electromechanical technologies have enabled the development of low power-consuming devices, efficient radio transceivings and unwise routing are still the main concerns in order to achieve improved network lifetime. Similarly, the problem of quality-of-service (QoS) and quality-of-experience (QoE) issues viz. video packets in WMSNs cannot be delayed more, otherwise success of the networks may be compromised. Wireless links are very reliable and can go down at any instance of time as they are comprised of low powered radios; hence, error control algorithms are required which must be energy efficient too. In addition to these, problems like scalability, security, and media access etc. are the common ones among the different types of sensor networks.

Moreover, all these aforementioned problems have a direct conflict with energy saving, which is the most limiting parameter for the WSN since when the nodes' batteries are depleted, the network is dead. Hence, energy efficiency is the most prominent problem in wireless sensor networks.

Also, the objectives or the mission of the network deployment must be taken care of, always with higher priority while devising solutions with regard to the aforementioned problems.

Here comes the cross-layer optimization into the picture to provide the most optimized solutions for the wireless sensor networks. Cross-layer indicates retrieving two or more parameters from different layers (from layered architecture of the network) and processing them in order to achieve more optimized solutions in WSNs. The primary goals of cross-layer optimizations in the sensor networks are to achieve power-efficient solutions, efficient routing, QoS and QoE provisioning, and optimal and energy-efficient scheduling in WSNs.

This chapter starts with a broader discussion on the layered architecture of wireless sensor networks and proceeds further with the discussion on the major drivers for cross-layer optimizations; then, numerous solutions for different layers will be discussed in its subsequent sections.

WSN Architecture and Protocol Stack

As already explained in Figure 16.1, sensor nodes are scattered/deployed in a sensing field to sense their environments. This sensed information is then transported towards the sink for further processing from where it can be accessed by the end user for his own purposes, as per the intended applications.

Moreover, in a wireless sensor network, sensor nodes are required to perform two different functions: *data origination* and *data routing*. Here, data origination refers to the ability of sensor nodes to generate data or measurements as per their nature, and data routing refers to the functionality of nodes where they act as the routers in order to forward packets to other nodes ultimately destined to the sink (i.e., multihop communication).

As in the traditional TCP/IP network, wireless sensor networks also employ five-layer protocol stack as depicted in Figure 16.2. However, some management planes are also there in the protocol stack to facilitate the sensors in managing power, movements, task distribution, as well as to coordinate the nodes while lowering the effective power consumption (Akyildiz et al., 2002).

The protocol stack consists of physical layer, data link layer, network layer[1], transport layer, and application layer. Along with these inherent layers in the protocol suit, there are a few planes to coordinate the nodes in their sensing events viz. power management plane, mobility management plane, and task management plane.

The power management plane controls the power consumption of the nodes in order to lower it so that the node can participate in the network actions for a longer duration, e.g., it may define when to keep radio on or off, when to broadcast in the network, and when to sense only, depending on the power level of the node. The mobility management plane controls the movement of nodes and also keeps a record of the nodes' movements. The task management plane controls the functioning of the node by defining the timing and degree of executable jobs in the sense

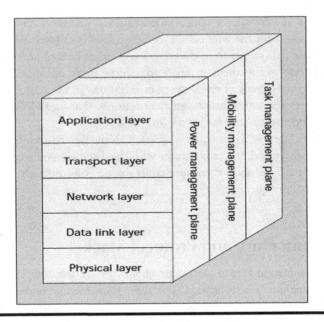

Figure 16.2 WSNs' protocol stack. (From Akyildiz et al., 2002.)

that it schedules the sensing of node or it decides what task a node can perform based on its power level.

Thus, through these planes, nodes in the network can work collaboratively in a power-efficient way.

Moreover, the roles and responsibilities of each layer in the protocol stack are as follows.

Physical Layer

The physical layer is responsible for a variety of lower-level tasks viz. frequency selection, carrier (frequency) generation, modulation, data encryption, etc. This layer is particularly responsible for working closely with the underlying hardware and transceiver design.

Data Link Layer

Like in TCP/IP protocol suit, the data link layer in WSNs is responsible for host-to-host communication. More illustratively, this layer is responsible for establishing point-to-point as well as point-to-multipoint connections in the network. The data link layer is mainly responsible for defining whom to allow to access to the medium, i.e., medium access and error control policies.

The medium access to the participating nodes is provided by first creating the necessary network infrastructures by creating the links for data transfer and then by efficiently sharing the network resources viz. energy, time, frequency etc. among the nodes in the network. A number of MAC protocols have already been proposed in the context of wireless sensor networks.

Another responsibility of the data link layer is to provide error control for the transmitted data. For this very purpose, there are two popular approaches: forward error correction (FEC) and automatic repeat request (ARQ). However, in the sensor networks with resource-constrained nodes, ARQ is limited by the retransmission cost and overhead whereas FEC heavily depends upon encoding and decoding complexity.

Network Layer

The network layer in sensor networks is responsible for routing of data packets from the originator to the sink. Due to their intrinsic nature, nodes in the sensor network are restricted to short range transmission only, i.e., they can only transmit to a limited distance, and hence direct communication between the source node and the sink may not always be possible, especially when they are distant. In such network scenarios, it may be required to utilize some intermediary nodes to carry the message to the sink on behalf of the source node, i.e., multihop communication. This functionality is the main responsibility of the network layer in the sensor network. Moreover, the network layer performs not only data routing but also implements data aggregation in order to achieve energy efficiency. Other than the data routing, internetworking with Internet, some relevant control system, and/or with some other sensor network is also taken care of by the network layer.

Transport Layer

Transport layer functionality is required at most in the cases when the network is planned to be accessed through the Internet or other external networks. The two major responsibilities of the transport layer are end-to-end reliability and congestion control. However, due to high energy cost and limited network resources, achieving end-to-end reliability is a little bit restricted, and localized reliability mechanisms are employed the most in sensor networks. Congestion control refers to the scenarios where high data traffic is generated by the events which may overwhelm the resource-constrained processing and storage capacity of the nodes; in addition, power consumption may be increased to the extent that nodes remain with no more energy to participate in the network actions.

Application Layer

The application layer mainly involves applications requesting data from the network and numerous management functionality.

Motivation behind the Evolution of Cross-Layer Optimization in WSNs

As explained earlier, the tiny size of nodes housing multiple modules (for sensing, transmitting, receiving, storing, processing, etc.) inside imposes a set of challenges in the network design using these nodes. The nodes become resource-constrained devices with limited storage, limited transmission, limited processing abilities. And, on top of these, limited power resource, which is very unlikely to be replaced because of the network's hostile deployment in most of its applications, is the most crucial one. Once the nodes start dying, network objectives start becoming compromised, leading to an unexpected end. Hence, energy can be stated to be the most valuable resource required to be preserved as long as possible.

Similarly, there are other limitations viz. redundancy due to dense deployment of nodes, lifetime of the network, energy-efficient routing, QoS and QoE issues in the mission-critical data, interference from other transmissions causing errors, network-capacity in terms of nodes deployed, hardware clock references for effective target tracking, security, etc. which are encountered while designing the network. All these aforementioned constraints are connected in the sense that providing a solution for one may have an impact on the other, and, also, each of the aforementioned limitations has a direct conflict with the most important requirement of lowering the energy consumption in the network. Their interconnections with one another and conflict with requirement of lowering the energy consumption can be easily intuited by the following discussion.

Redundancy: Redundancy in data refers to reporting of similar data from multiple nodes, especially when they are located in a close vicinity, i.e., when they are placed close enough to one another. Nodes placed/deployed in this manner may generate highly correlated or even repeated data which is treated as the wastage of the most precious energy resource.

Network Lifetime: Definition of the network lifetime must be application-specific. For example, if it is defined as the time when the first node dies in the network, it won't find its applicability in the applications which can survive, even with a number of packet losses or the ability to tolerate nodes' failure to some extent. Similarly, when it is defined as the time when all the nodes die in the network, applications delivering mission-critical data may suffer with an unrecoverable loss.

Routing: A routing protocol is required when a source node needs some intermediary nodes in getting its data delivered to the sink. In an attempt to establish path, flooding of control packets may be executed, which clearly indicates a huge consumption of energy. Similarly, in view of changing network topology, route re-computation results in energy consumption. In addition to these, constant control packet exchange for route maintenance may further consume the nodes' energy. Thus, an energy-efficient routing with the least possible control overhead is required at most in the sensor network.

QoS and QoE Issues: Quality of service and quality of experience issues are the most prominent ones when applications deployed for the sensor network yields mission-critical data. For example, while expecting multimedia data for surveillance purposes and health monitoring data from a deployed WBAN, any delay incurred may result in an unrecoverable loss and may lead to the network failure. In order to ensure timely delivery of such critical data, it may be required to have frequent access to the shared medium and high power transmission, which definitely contradicts the theory of lowering power consumption.

Error Control: Due to the inherent nature of wireless transmission, it is very likely to encounter nearby interferences, which result in data-errors, i.e., either erroneous packets will be transmitted or packets may get lost in the network. In this regard, two popular approaches are available, namely, FEC and ARQ. However, in the sensor networks with resource-constrained nodes, ARQ is limited by the retransmission cost and overhead, whereas FEC heavily depends upon encoding and decoding complexity. Moreover, both the approaches require high power consumption.

Network Capacity: Network capacity is another issue which if not handled properly results in unwise power consumption. More illustratively, if fewer numbers of nodes are deployed into the network, it is very likely that nodes would be performing high power communication to the other nodes in order to achieve the network objective. In some cases, it might be very possible that some nodes deployed at the boundary of the field become isolated as the intermediary nodes or relay nodes are not left with suitable amount of energy. Similarly, when a large number of nodes are deployed in the sensing area, although coverage increases at comparatively low energy cost, the network suffers with congestion and energy-efficient routing arises as a big challenge.

Hardware Clock References: Applications like target detection or tracking are heavily dependent on the sensor's clock references, as the event timing along with the position might be required for a successful detection. Here, synchronization accuracy of the clock requires more energy consumption.

Security: Since the very nature of communication medium in sensor network is the broadcast, nodes are more vulnerable to the security threats as they may suffer with a number of attacks viz. spoofing attacks, wormhole attacks, Sybil attacks, etc. More illustratively, in the spoofing attack, the malicious nodes join the network by stealing a valid address of the participating node via getting through the packets broadcasted in the air. Then, the malicious nodes start transmitting false packets using the stolen address. Wormhole attack indicates that the malicious nodes capture data from one part of the network and forward the captured data into another portion of the network through a different communication link. Similarly, in the Sybil attack, the attacker assumes multiple identifications and disrupts the network functioning by broadcasting a number of interfering packets.

As solutions to such attacks, cryptography can be used, but the solutions' requirement of exchanging additional bits for encryption and decryption processes increases the overall energy consumption in the network, hence conflicting with requirement of lowering the energy consumption.

Therefore, from the discussion provided above, it can be concluded that the aforementioned limitations are interconnected and are directly conflicting with the prime requirement of reducing the network energy consumption and hence a comprehensive study is mandatory to identify the tradeoffs aiming the requirements. Moreover, a cross-layer solution is required while dealing with these interdependent problems in order to achieve the ultimate objective of improved network lifetime.

The idea cross-layer solutions or optimizations involves two or more layers (in the sensors' protocol stack) which may or may not be adjacent to each other. For example, data link layer might be informed about the quality of service requirements to achieve a better scheduling of nodes for the ongoing applications; similarly, the details of the channels' current status can be made available to the network layer in order to restrict the usage of bad channels in the routing paths as shown in Figure 16.3.

Thus, a cross-layer optimization can be seen as the message exchange or interaction between two or more adjacent/non-adjacent layers in order to achieve the optimal solution while confirming the tradeoffs among the various application-specific requirements.

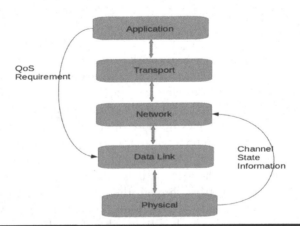

Figure 16.3   An instance of cross-layer interaction.

Cross-Layer Protocols

This section focuses on the various cross-layer solutions presented in the context of wireless sensor networks at each layer as per the protocol stack.

Application Layer

The significance of cross-layer optimization has been demonstrated in an attempt to improve the lifetime of sensor network with clustered network architecture in Yuan et al. (2006). In the solution provided, QoS measures were taken into consideration by the adaptive code position modulation scheme (ACPMS) (Yuan et al., 2005) at the cluster heads, and the transmission are adapted in accordance with the computed bit error rate (BER). Simulations have established that the cross-layer scheme proposed in Yuan et al. (2006) achieves network longevity by 60 rounds (here, rounds are the instances of cluster reformation) while providing QoS in terms of delivered packet ratio and end-to-end delay.

Another instance of such a solution in this category at the application layer is the one provided in Chen et al. (2007) through a cross-layer routing scheme entitled directional geographical routing (DGR) for transmission videos using H.26L standard. H.26 along with MPEG are the two most popular standards for the video compression. Video compressed using these are so tightly packed that they require less bandwidth for video transmission. DGR decomposes the long video stream into a number of small flows, each being transmitted through different routes and hence avoiding the single route transmission which causes more energy consumption. Moreover, FEC is employed to guarantee the arrival of entire video stream.

In Liang et al. (2007), channel state information (CSI) is used to schedule transmission only to the sensor experiencing better channels, hence enhancing the data fusion and improving the network lifetime.

Other than these, several other cross-layer schemes at the application layer have been proposed in Al-Karaki et al. (2009), Chen et al. (2006a), Wang et al. (2008, 2009), Akan and Akyildiz (2005), and Galluccio et al. (2002).

Transport Layer

In order to alleviate congestion in the wireless sensor networks, transport layer functionality, when considered along with that of the network layer, a scheme, biased geographical routing (BGR) (Popa et al., 2006) came into existence while exploring the on-demand multipath routing mechanism. BGR splits the data traffic in the event of network congestion. However, since the bias (determining the extent to which split traffic will deviate from the original path) is chosen randomly, the scheme may perform poorly in some scenarios. Similarly, the congestion aware routing (CAR) (Kumar et al., 2008) is another example in this category which implements priority-aware routing in order to alleviate network congestion and proposes multiple sinks to gather data packets with different priorities.

Similarly, in Chiang (2005), there comes a cross-layer solution for congestion control and power control as it brings transport layer along with physical layer.

Network Layer

Routing being the main function of the network layer has attracted a lot of attention of the researcher community as it has a great impact over the network lifetime.

An extension of directed diffusion was proposed as energy-efficient differentiated directed diffusion (EDDD) (Chen et al., 2006b), which discriminates between the real-time (RT) and best effort (BE) flows in order to route them accordingly by considering the time constraints for RT and energy efficiency for BE.

In Wang et al. (2007), a cross-layer priority-based congestion control protocol (PCCP), which introduces two different queues between the network and data link layer, is proposed for WSNs. The first queue is meant for the packets generated by the nodes whereas the second queue holds the traffic routed through the node. PCCP figure out the congestion on the basis of analysis made over the inter-arrival times of packets to be routed through. In each case, it determines congestion; it floats an implicit congestion notification (ICN) packet to the nodes to lessen the speed of message forwarding, hence reducing the network congestion.

Similarly, in Nam et al. (2007), a cross-layer protocol has been proposed to achieve a tradeoff between delay and energy consumption. The protocol proposes a new adaptive MAC (A-MAC), which introduces a new metric depending upon the duty cycle to modify the routing decision.

Moreover, several other routing-based cross-layer solutions have been proposed in the literature, emphasizing achieving energy efficiency or some other application-specific tradeoffs by modifying one or more parameters from other layers along with the one(s) from network layer.

Data Link Layer

Among the two main responsibilities of data link layer, i.e., medium access and error control, medium access has attracted a lot to work upon. Especially in the cross-layer optimization techniques in wireless sensor networks, medium access control has made its presence almost in every proposal, if not explicitly, then at least implicitly.

In Lin and Kwok (2006), a cross-layer energy-efficient approach is proposed under the title, channel adaptive approach to energy management (CAEM), which implements carrier-sense multiple access with collision detection (CSMA–CD) as its media access protocol. A node continuously senses the medium, and whenever it has something to send, it proceeds only if it finds a channel in good health;

otherwise, it backtracks and goes into sleep mode. Through the application of fair scheduling and queuing algorithms, starvation is avoided if the channel resumes after a long time. It has been shown that the scheme improves the network lifetime by 40% when compared to the original low-energy adaptive clustering hierarchy (LEACH) (Heinzelman et al., 2000) through the knowledge of channel states.

In Kwon et al. (2006), using time-division multiple access (TDMA) as its medium access protocol, a cross-layer routing protocol has been proposed which also implements automatic repeat request for error control. The algorithm verifies channel states at every link in order to determine the best route for each message stream along with defining adaptively the retransmission limit to increase the probability of successful packet delivery.

In Chen et al. (2007), two physical layer parameters—channel state information and residual energy information—have been proposed for the consideration, along with a MAC protocol. In the scheme, nodes are scheduled for transmission as per their sensing of healthy channels when they are with full energy, and when energy in the nodes reduces, nodes are scheduled for transmission in the order of their residual energies.

In fact, the literature is filled with schemes which exploit data link layer parameters along with those of other layer in order to achieve energy-efficient operations suited to the requirements of concerned applications.

Physical Layer

Some of the algorithms which utilize the physical layer parameters in achieving the cross-layer solutions have already been discussed in previous sections. In addition to those, some are being listed as follows.

In Tian and Ekici (2007), multihop task mapping and scheduling (MTMS), which looks for the best application task mapping at each node, has been proposed.

In Liang and Liang (2009), virtual multiple in, multiple out (MIMO) communication-based (through the collaborative efforts of neighbor nodes in close vicinity) cross-layer approach has been proposed in order to achieve energy-efficient routing in the network.

The aforementioned solutions using cross-layer optimization are just some of the solutions. In addition to these, there are numerous other protocols falling in this category which could be explored by the readers.

Cross-layering has evolved as a great solution paradigm which could be explored for providing solutions for a variety of applications depending upon the different types of sensor networks. It provisions a number of different aspects from different layers to be included in the solution, e.g., QoS provisioning of application layer may be provided for the consideration to the MAC layer in order to schedule access to the medium; MAC can be facilitated with the physical layer specifications which can further be used by network layer devising routing path to the sink, etc.

The most ideal cross-layer solution is the one which addresses the issues from each layer of the protocol stack. However, it requires a lot of effort in designing a comprehensive cross-layer solution which can address the issues of each layer in the protocol-suite in an energy-efficient manner.

End Note

1. Some authors do not consider the transport layer as a part of the protocol stack as it has been found to increase the complexity and waste the sensors' energy.

References

O. B. Akan and I. F. Akyildiz, Event-to-sink reliable transport in wireless sensor networks. *IEEE Transactions on Networking*, vol. 13, no. 5, pp. 1003–1016, 2005.

I. F. Akyildiz, W. Su, Y. Sankarasubramaniam and E. Cayirci, A survey on sensor networks. *IEEE Communications Magazine*, vol. 40, pp. 102–114, 2002.

J. N. Al-Karaki, R. Ul-Mustafa and A. E. Kamal, Data aggregation and routing in wireless sensor networks: Optimal and heuristic algorithms. *Computer Networks*, vol. 53, no. 7, pp. 945–960, 2009.

M. Chen, T. Kwon and Y. Choi, Energy-efficient differentiated directed diffusion (EDDD) in wireless sensor networks. *Computer Communications*, vol. 29, no. 2, pp. 231–245, 2006b.

M. Chen, V. C. M. Leung, S. Mao and Y. Yuan, Directional geographical routing for real-time video communications in wireless sensor networks. *Computer Communications*, vol. 30, no. 17, pp. 3368–3383, 2007.

X. Chen, Y. Xiao, Y. Cai, J. Lu and Z. Zhou, An energy diffserv and application-aware MAC scheduling for VBR streaming video in the IEEE802.15.3 high-rate wireless personal area networks. *Computer Communications*, vol. 29, no. 17, pp. 3516–3526, 2006a.

Y. Chen and Q. Zhao, An integrated approach to energy-aware medium access for wireless sensor networks. *IEEE Transactions on Signal Processing*, vol. 55, no. 7, pp. 3429–3444, 2007.

M. Chiang, Balancing transport and physical layers in wireless multihop networks: Jointly optimal congestion control and power control. *IEEE Journal on Selected Areas in Communications*, vol. 23, no. 1, pp. 104–116, 2005.

L. Galluccio, A. Campbell and S. Palazzo, CONCERT: Aggregation-based congestion control for sensor networks. In: *Proceedings of the 3rd ACM SENSYS Conference*, 2002, pp. 274–275.

W. B. Heinzelman, A. P. Chandrakasan and H. Balakrishnan, Energy-efficient communication protocol for wireless microsensor networks. In: *Hawaii International Conference on System Sciences (HICSS)*, 2000, pp. 10–19.

R. Kumar, R. Crepaldi, H. Rowaihy, A. F. Harris, G. H. Cao, M. Zorzi and L. T. Porta, Mitigating performance degradation in congested sensor networks. *IEEE Transactions on Mobile Computing*, vol. 6, no. 6, 2008, pp. 682–697.

H. Kwon, T. H. Kim, S. Choi and B. G. Lee, Across-layer strategy for energy-efficient reliable delivery in wireless sensor networks. *IEEE Transactions on Wireless Communications*, vol. 5, no. 12, pp. 3689–3699, 2006.

J. Liang and Q. Liang, Channel selection in virtual MIMO wireless sensor networks. *IEEE Transactions on Vehicular Technology*, vol. 58, no. 5, pp. 2249–2257, 2009.

Q. Liang, D. Yuan, Y. Wang, H. Chen. A cross-layer transmission scheduling scheme for wireless sensor networks. *Computer Communications*, 30 (14–15): 2987–2994, 2007.

X.-H. Lin and Y.-K. Kwok, CAEM: A channel adaptive approach to energy management for wireless sensor networks. *Computer Communications*, vol. 29, no. 17, 3343–3353, 2006.

Y. Nam, T. Kwon T, H. Lee, H. Jung, Y. Choi. Guaranteeing the network lifetime in wireless sensor networks: a MAC layer approach. *Computer Communications*, 30(13): 2532–2545, 2007.

L. Popa, C. Raiciu, I. Stoica and D. Rosenblum, Reducing congestion effects in wireless networks by wireless networks by multipath routing. In: *Proceedings of the 14th IEEE ICNP Conference*, 2006, pp. 96–105.

Y. Tian and E. Ekici, Cross-layer collaborative in-network processing in multihop wireless sensor networks. *IEEE Transactions on Mobile Computing*, vol. 6, no. 3, pp. 297–310, 2007.

C. Wang, B. Li, K. Sohraby, M. Daneshmand and Y. Hu, Upstream congestion control in wireless sensor networks through cross-layer optimization. *IEEE Journal on Selected Areas in Communications*, vol. 25, no. 4, pp. 786–795, 2007.

H. Wang, D. Peng, W. Wang, H. Sharif and H.-H. Chen, Cross-layer routing optimization in multirate wireless sensor networks for distributed source coding-based applications. *IEEE Transactions on Wireless Communications*, vol. 7, no. 10, pp. 3999–4009, 2008.

W. Wang, D. Peng, H. Wang, H. Sharif and H.-H. Chen, Cross-layer multirate interaction with distributed source coding in wireless sensor networks. *IEEE Transactions on Wireless Communications*, vol. 8, no. 2, pp. 787–795, 2009.

Y. Yuan, Z. Yang, J. He and W. Chen, An adaptive code position modulation scheme for wireless sensor networks. *IEEE Communication Letters*, vol. 9, no. 6, pp. 481–483, 2005.

Y. Yuan, Z. Yang, Z. He and J. He, An integrated energy aware wireless transmission system for QoS provisioning in wireless sensor network. *Computer Communications*, vol. 29, no. 2, pp. 162–172, 2006.

Chapter 17

Cross-Layer Quality of Service Approaches in Wireless Sensor Networks

Wireless sensor networks (WSNs) have evolved as ones which bring automated sensing, embedded computing, and wireless networking together in an integrated way such that a set of small electromechanical devices (particularly, sensor nodes) sense its surroundings in order to generate some environmental data which can be forwarded to some centralized processing station (sink or base station) either to take some corresponding course of action or for further user access as per the nature of deployed applications. The initial deployment of the wireless sensor network was made for low data rate and non-real-time applications such as agriculture and environmental monitoring, but due to the huge technological advancements, the deployment of more capable nodes, e.g., camera sensors, audio sensors, etc., has enabled new applications viz. smart grid, intelligent transportation system, industrial monitoring, wireless body area networks (WBANs), etc. Such applications require certain Quality of Service (QoS) to be ensured instead of just having best-effort performance as in its initial deployments for traditional applications.

More illustratively, when WSNs are deployed for power grid infrastructure towards achieving smart grid implementation, various QoS requirements, e.g., latency, reliability, security, spectrum availability, are to be considered in the concerned problems of teleprotection systems, emergency power restoration, and substation monitoring and control.

In the applications of military and security prospects, when deployed, wireless sensor networks are required to provide real-time services with the full reliability guarantee; otherwise, missions and the lives of soldiers/civilians would be compromised.

In healthcare and environmental monitoring applications such as patient monitoring, wireless body area networks, home health care monitoring for elderly and chronic patients, and ecological monitoring applications, ensuring QoS is a strict requirement, as delayed reporting may prevent from taking cautious time-bound actions.

In industrial applications, e.g., pipeline monitoring, nuclear power plant monitoring, etc., WSNs have emerged as a very popular choice, due to low cost, rapid deployment, self-configuration capability, and low power consumption, to replace manual diagnoses. Timely and reliable delivery of delay-critical data in such applications is a mandatory QoS requirement which may put a huge impact over the success of application.

Similarly, in order to achieve intelligent transportation system (ITS) through the deployment of sensor networks, efforts must be made to provide delay- and reliability-aware solutions for the safety-related and traffic management and control applications.

In addition to the aforementioned applications, there are numerous other applications in which assuring QoS is a mandatory requirement. However, when it comes to providing a generic definition of QoS, there is no such consensus achieved yet and the definition of QoS depends upon the applications concerned. For example, QoS represents the service requirements of a particular application which are to be confirmed by the network. Similarly, it also refers to the ability of the network to adapt to the specific class of data. Moreover, the International Telecommunication Union (ITU) Recommendation E.800 (09/08) defines QoS as: "Totality of characteristics of a telecommunication service that bear on its ability to satisfy stated and implied needs of the user of the service."

In the simplest way, QoS can be described as defining different priorities for the different applications, users, and data flows on the basis of various application-specific requirements by controlling the resource sharing (Yigitel et al., 2011). Moreover, the QoS can also be referred to as the ability to satisfy the needs of the user or application as shown in Figure 17.1. From Figure 17.1, it can be easily concluded that without being concerned how the network would manage its resources in order to provide the QoS support, the application or user puts application-oriented requests over the network.

QoS provisioning is classified into two classes: hard QoS and soft QoS (Anbagi et al., 2016). *Hard QoS* specifies strict compliance with the metrics defining the application or user requirements, i.e., providing the guaranteed QoS, whereas *soft QoS* signifies tight QoS requirements but with limited flexibility, i.e., QoS provisioning can tolerate some temporal violation to some extent (Tsigkas and Pavudou, 2008).

In order to provide hard QoS and soft QoS, there are two service differentiation models which are widely used in the wireless networks (in wired networks too)

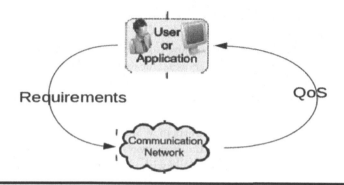

Figure 17.1 QoS model.

viz. *integrated service* and *differentiated service*; however, both the models proceed with assigning suitable priorities to the *flows* or *packets*, mapping the priorities into service qualities, i.e., quantifying the priorities, and then provide the requested service quality through resource sharing. An instance of the same can be observed in Figure 17.2.

More illustratively, integrated service, also termed IntServ, is basically a resource reservation-based approach which maintains services on per flow basis such where a flow is either data-centric or host-centric. A data-centric flow is a data packet generated by event-triggered sensor nodes in response to the occurrence of some events that nodes are configured to detect, whereas a host-centric flow refers to the data packets being exchanged between host and destination. As a matter of fact, providing guaranteed services in wireless environments is quite difficult, primarily due to time-varying channel capacity. Other than that, as there are a huge number of nodes in wireless sensor networks, maintaining per flow states is a very tough

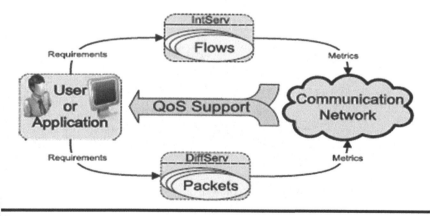

Figure 17.2 QoS provisioning using integrated and differentiated service model. (From Yigitel et al., 2011.)

practice; also, providing reliable QoS signaling in order to facilitate resource sharing, as per the integrated service approach, is quite complex.

Differentiated service, popularly known as DiffServ, is a reservation-less model which provisions per-packet-based services. Due to its simplicity and easy-to-implement property, the DiffServ model can be adapted to the wireless sensor network; however, its high cost-of-memory requirement is treated as a major problem of this approach.

Providing end-to-end QoS supports in WSN is a major requirement for the success of the network deployed and has always been a topic of great interest for the researcher community, especially with changing network scenarios after the huge technological advancements. A large number of mechanisms and algorithms in different layers of protocol stack have already been proposed in this regard to provide QoS supports. In this chapter, focus will be made over the cross-layer approaches for providing QoS supports in the wireless sensor network. As explained in previous chapter, cross-layer approaches involve multiple parameters from the different layers in sensors' protocol stack. These schemes provision communication between two or more layers, as demonstrated in Figure 17.3, depending on the requirements specification of the concerned application.

Moreover, characteristics of different networks also affect the QoS provisioning in the network a lot, e.g., in the context of mobile sensor networks (MSNs), frequently changing network topology, and, hence, the instability of network

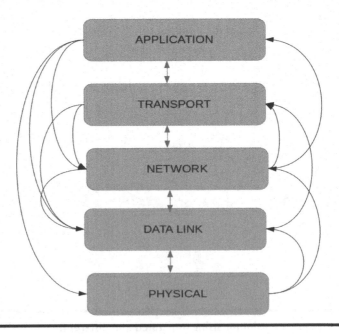

Figure 17.3   Cross-layer communication.

links makes it quite difficult to provide QoS support. Other than the network characteristics, maintaining application-specific QoS requirements is also a very challenging issue, as with the evolution of new technologically rich applications, existing QoS parameters may be inefficient and, hence, new parameters may be required to provision. In the subsequent sections, a discussion will be made on the different QoS perspectives followed by the discussion over the various QoS challenges in WSN.

QoS Perspectives in WSN

Perspective simply refers to the QoS aspect of interest. In view of discriminating how the underlying network provides QoS supports for the applications deployed, there are two different perspectives of QoS in wireless sensor networks: application-specific and network-specific perspectives (Chen and Varshney, 2004).

Application-Specific QoS Perspective

The application-specific perspective of QoS represents the specific requirements of the application being deployed. For example, the applications like targeted monitoring may impose specific requirements over the deployment of nodes in order to provide a prescribed degree of coverage (Meguerdichian et al., 2001) or specific requirements on measurement precision and the number of active sensors in order to limit measurement errors (Iyer and Kleinrock, 2003).

Network-Specific QoS Perspective

The network-specific QoS perspective is concerned with how the QoS-constrained data would be delivered to user/sink through an effective utilization of network resources. Some of the network-specific requirements imposed by the data to be delivered are delay or latency, reliability, packet loss, etc.

QoS Challenges in WSN

In the implementation of QoS mechanisms, traditional wireless networks confront a number of challenges, e.g., the problem of time-varying channel capacity and unreliable communication links. In addition to these, due to being a variant of challenged networks, wireless sensor networks also face the problems of limited resources, asymmetric data transfer, long delays, and high error rate. These aforementioned sufferings, when combined with those of the traditional wireless networks, result in a number of QoS challenges, some of which are being mentioned as follows:

Resource Constraints

The wireless sensor network is a resource-constrained network in the sense that its constituent nodes come up with limited sensing capacity, limited transmitting and receiving ability, limited computational capabilities, and limited storage capacity. And, above all these aforementioned limitations, power constraint is the most severe one. Due to its tiny size, only small batteries can be installed within the nodes which are mostly non-replaceable and/or non-replenishable due to the hostile nature of the surroundings of deployment. Therefore, in order to provide a solution with requested QoS supports, computationally complex and energy-consuming algorithms must be avoided.

Node Deployment

Nodes in the WSNs can be deployed either deterministically or randomly. In a case where nodes are deployed in a deterministic way, providing QoS support becomes a comparatively easy task as the nodes can be manually installed/deployed in the network and the routing paths can be easily controlled. However, in a case where the deployment strategy being followed is random, nodes are supposed to organize themselves on their own; hence, requirements like neighbor discovery, routing path determination, etc. are significant issues in order to provide required QoS supports.

Dynamic Network Topology

Topology refers to the geometrical layout of the nodes in network. In wireless sensor networks, there are a number of factors that can alter the topology of the network viz. battery depletion of the nodes, sudden malfunctioning of the nodes, nodes' movement, natural disastrous situation like flood, link failures, etc. Therefore, maintaining QoS support for such dynamic network induces a great challenge.

Data Redundancy

A wireless sensor network can be seen as a huge collection of nodes deployed densely to sense and report the environment. This dense deployment of such a large number of sensor nodes results in redundancy, i.e., high spatiotemporal correlated data. Though redundancy enhances the reliability of network, it also causes unnecessary transmission of duplicate data. However, this redundancy can be taken care of by some data aggregation schemes, but they, in turn, induce additional delay and complexity in data delivery in the network. Therefore, while providing QoS support in the network, data redundancy must also be dealt with accordingly.

Traffic Asymmetry

Depending on the nature of application deployed, data flow in the network can either be periodic or aperiodic. For example, in case of normal monitoring applications, data flow is usually periodic whereas in the event-triggered sensor applications, it is aperiodic, i.e., a number of nodes may simultaneously report in the presence of some event. Therefore, such asymmetric traffic must be brought under consideration while formulizing QoS support in the network.

Scalability

A wireless sensor network may comprise a very large number of sensor nodes; therefore, the formulation of QoS support must also be scalable to cope with this large number of nodes.

Energy Efficiency

Energy efficiency is the most significant aspect of a healthy network design, as the nodes in the network suffer with power-scarcity. Once depleted, power units are hard to replace/replenish. Hence, the QoS mechanism being proposed must establish an application-suiting tradeoff between the energy consumption and other requirements of application. More illustratively, while ensuring Quality of Service, residual energy of the nodes must be provided with adequate attention.

Traffic Heterogeneity

Nowadays, nodes with different abilities are being deployed in the sensing field, e.g., in the ecological monitoring application, multiple nodes can be deployed with the capability of temperature sensing, audio sensing, image sensing, etc. Nodes with different abilities may generate different kinds of data required to be transported to the sink, and, therefore, proper treatment must be given while proposing QoS mechanism in such applications.

Sink Multiplicity

In some applications of sensor networks, more than one sink/base station might be required to be deployed in different regions of a wide sensing field. In such network scenarios, multiple sinks may result in different data with different directions, and handling these different streams of data while providing necessary sets of QoS supports in the network sounds to be very challenging.

High Error Rate

The wireless network deployed using sensor nodes suffers with a high bit error rate (BER) as the communication links fail very frequently for a number of reasons. Therefore, QoS mechanisms with suitable error control methods must be devised in WSNs.

Extreme Network Environment

Based on the very nature of applications, nodes in WSNs are deployed in the terrestrial environment, underwater environment, and underground environment. When nodes are deployed underwater, they may face unexpected and unrecoverable displacement due to flood or tsunami, or they may even get destroyed due to heavy waves, etc. Similarly, when they are buried underground, they are more prone to corrosion; and, lastly, when they are thrown into some human-inaccessible area in the terrestrial environment, they are subjected to unexpected damage. While provisioning QoS mechanisms for such networks, the aforementioned severe environment conditions must be taken care of adequately.

Real-Time Traffic

In the mission-critical applications viz. disaster management and recovery or security surveillance, nodes are supposed to generate real-time data which require time-bound responses, as the data generated from such applications might become invalid after a very short time duration. Providing QoS supports in these applications with delay-sensitive and reliability-sensitive data is a quite challenging task.

Security

Like in traditional wireless networks, pursuance of security aspects in WSNs is a very important job for the successful implementation of network. Due to the aforementioned challenges viz. the resource-constrained nature of nodes, etc., security techniques of the traditional wireless networks cannot be directly employed in wireless sensor networks. Any such security technique for WSN must bring resource constraints (like limited processing, sensing, communication, and storage capacity along with the limited non-replenishable battery unit) in addition to the other above mentioned challenges along with QoS requirements like delay and reliability in its consideration.

Cross-Layer QoS Approaches in WSN

QoS provisioning refers to providing the desired level of Quality of Service to the user of the concerned application either in (single)-layered or cross-layered protocol while maintaining the overall network parameters unchanged. In the traditional layered approach, the focus is to optimize the protocols in the individual layer in order to achieve the desired QoS provisioning in a particular layer, whereas the cross-layered approaches provide the desired QoS provisioning by optimizing the interaction among two or more layer protocols, i.e., the cross-layer approaches process the parameters and metrics associated with different protocol layers—physical, data link, network, transport, and application—altogether to obtain the desired level of QoS provisioning.

Cross-layer approaches achieve the optimization of overall performance of the wireless sensor network along with provisioning the QoS as requested by the user/ application; however, the major disadvantage of such approaches is the strong interdependence or tight coupling among the different layers of protocol suite. More illustratively, in cases where changes are required to be incorporated into a single layer, all the other layers with which it had interacted previously are also required to be tuned accordingly. But, in view of achieving the desired level of QoS supports viz. timeliness and reliability in mission-critical monitoring applications, cross-layer approaches have proved their significance.

In the following sections, the popular cross-layer contributions will be discussed in order of their interactions among the layers in protocol stack.

Two Layers' Interaction

In Bhuiyan et al. (2011), an interaction between the network layer and the application layer has been demonstrated in order to provision timeliness and reliability of data being transmitted. In this work, a routing protocol has been proposed which not only avoids congestion but also reduces the overall delay in the transmission of data via selecting a node with a light load which is expected cause less delay in forwarding the data packets towards the sink. The success of the scheme heavily depends on the periodic broadcasting of control packets which are comprised of congestion status of the nodes along with their respective delay measurements; here, the delay measurement refers to the estimation of end-to-end delay in delivering the packet to the base station by the nodes. The network-wide sharing of such control packets enables the neighbor nodes to use the broadcasted data in the route selection process. Moreover, a congestion mitigation method has also been implemented in the scheme through the active participation of MAC layer and application layer. MAC layer always sends feedback about its achievable data forwarding rate to the network layer, and the application layer is instructed by the protocol when to decrease its traffic generation rate. A fraction of the packets received from the other nodes are dropped by the network layer if the incoming rate is higher than the data forwarding rate reported by the MAC layer. Thus, through such congestion avoidance and mitigation techniques, timeliness is provisioned in the scheme.

In Ababneh et al. (2012), cross-layering has been implemented via an interaction between network and physical layers in order to devise a reliability-aware routing solution. The problem of routing, along with that of bandwidth allocation and flow assignment in the WBANs, has been investigated as a linear optimization problem. The scheme defines a metric with the name *network utility* which considers both the nodes' residual energy and the throughput for establishing the routing path and adaptively allocating bandwidth to the nodes in the network.

Similarly, in Yuan et al. (2006), cross-layering involves parameters from the application layer and physical layer to develop an integrated adaptive wireless transmission system for wireless sensor networks which aims at QoS requirements viz. reliability and timeliness in an energy-efficient way. The scheme analyzes the QoS requirements from the application layer together with the modulation and transmission techniques from the physical layer while integrating them into a single framework where the framework employs a clustering-based protocol with cluster heads (CHs) forming a multihop backbone for communication. On the basis of analysis results of both inter-cluster and intra-cluster QoS performances, they have modeled the overall network-wide energy consumption and QoS in terms of BER performance, end-to-end transmission latency, and packet loss ratio by the communication parameters of CHs, including the transmit power and modulation level (bits/symbol). The scheme proposes a centralized off-line protocol to adjust the communication parameters of CHs for the fixed QoS provisioning, along with the proposal of a distributed online protocol aiming at dynamic QoS provisioning.

Three Layers' Interaction

In Choe et al. (2009), cross-layering has been demonstrated by including parameters from application, network, and data link layers to achieve the expected information quality at the end system via controlling of data reporting function. The scheme comprises a QoS-aware data reporting tree construction (QRT) scheme and QoS aware node scheduling (QNS) scheme. In the proposed scheme, QRT constructs a data reporting tree on the basis of conditions of end-to-end delay and traffic load to figure out the paths from the cluster heads to the sink, whereas QNS schedules a definite number of nodes based on QoS requirements in the cluster to report to their respective cluster heads in a collision free manner.

In Wang et al. (2010), an interaction among the application layer, MAC layer, and physical layer has been demonstrated via proposal of a hybrid MAC protocol aimed at providing QoS supports for delay-sensitive traffic flows. The proposal combines channel-allocation schemes from the existing contention-based and TDMA-based protocols to implement the tradeoffs between different performance metrics. The scheme comprises a channel-reservation technique which reduces the end-to-end delay for the latency-sensitive traffic flows via provisioning them to pass through multiple hops within a single MAC frame with higher priority in order to reduce probable queuing delay.

Four Layers' Interaction

Francesco et al. (2011) have performed the cross-layering of application, transport, network, and MAC layers in order to formulize a reliable and energy-efficient data-collection protocol for IEEE 802.15.4-based wireless sensor networks. In their proposal, an adaptive and cross-layer framework for reliable and energy-efficient data collection in WSNs has been implemented which involves an energy-aware adaptation module capturing the application's reliability requirements, and configuring the MAC layer based on the network topology and the traffic conditions with an objective to minimize the energy consumption. In particular, a low-complexity distributed algorithm, called ADaptive Access Parameters Tuning (ADAPT) has been proposed which effectively achieves the application-based QoS support for reliability under a wide range of operating conditions in both networking scenarios: single-hop and multi-hop.

In Park et al. (2011), cross-layering among the application layer, network layer, MAC layer, and physical layer has been implemented in order to achieve the QoS support for the reliability and timeliness in the resource-constrained wireless sensor networks. The scheme proposes a novel protocol entitled Breath for the control applications which focuses on the networks facilitating multihop routing among the nodes and sink under certain reliability and latency requirements. Randomized routing, carrier-sense multiple access with collision avoidance (CSMA/CA), and randomized sleep–awake discipline are optimized to minimize the power consumption in the scheme. The scheme establishes relations among the reliability, latency, and total network energy consumption as a function of routing, MAC, physical layer, duty cycles, and radio powers. The design approach is based on a constrained optimization problem, whereby the objective function is the energy consumption and the constraints are the packet reliability and delay.

In addition to the aforementioned schemes, a number of cross-layer protocols targeting providing the various QoS supports for the applications of interest exist in the literature, establishing interactions among a number of layers in the concerned protocol stack.

Although the design of new cross-layer QoS approaches faces a number of challenges, such as limited resources, requirement of providing consistent QoS over time, support for heterogeneous data traffic, support for mobility, etc., the emergence of WSNs as the most promising platform for the development of a wide variety of applications and the benefits in terms of overall network optimization using the cross-layered approaches are consistently compelling the researchers to devise new cross-layer solutions meeting the application-specific QoS requirements viz. reliable and timely delivery of data in mission-critical and real-time applications, etc.

References

N. Ababneh, N. Timmons and J. Morrison, Cross-layer optimization protocol for guaranteed data streaming over wireless body area networks. *Proceedings of the 8th IWCMC*, 2012, pp. 118–123.

I. AI-Anbagi, M. Erol-Kantarci and H. T. Mouftah, A survey on cross-layer quality-of-service approaches in WSNs for delay and reliability-aware applications. *IEEE Communication Surveys & Tutorials*, vol. 18, no. 1, pp. 525–552, First Quarter 2016.

M. Aykut Yigitel, O. D. Incel and C. Ersoy, QoS-aware MAC protocol for wireless sensor networks: A survey. *Computer Networks*, vol. 55, pp. 1982–2004, 2011.

M. M. Bhuiyan, I. Gondal and J. Kamruzzaman, CODAR: Congestion and delay aware routing to detect time critical events in WSNs. *Proceedings of the IEEE ICOIN*, January 2011, pp. 357–362.

D. Chen and P. K. Varshney, QoS support in wireless sensor networks: A survey. In: *Proceedings of the 2004 International Conference on Wireless Networks (ICWN 2004)*, 2004, pp. 227–233.

H. J. Choe, P. Ghosh and S. K. Das, Cross-layer design for adaptive data reporting in wireless sensor networks. *Proceedings of the IEEE International Conference on PerCom*, March 2009, pp. 1–6.

M. Di Francesco, G. Anastasi, M. Conti, S. K. Das and V. Neri, Reliability and energy-efficiency in IEEE 802.15.4/Zigbee sensor networks: An adaptive and cross-layer approach. *IEEE Journal on Selected Areas in Communications*, vol. 29, no. 8, pp. 1508–1524, 2011.

R. Iyer and L. Kleinrock, QoS control for sensor networks. *ICC* 2003, May 2003.

S. Meguerdichian, F. Koushanfar, M. Potkonjak and M. B. Srivastava, Coverage problems in wireless ad-hoc sensor networks. In: *Proceedings of IEEE Infocom*, 2001, pp. 1380–1387.

P. Park, C. Fischione, A. Bonivento, K. H. Johansson and A. Sangiovanni-Vincent, Breath: An adaptive protocol for industrial control applications using wireless sensor networks. *IEEE Transactions on Mobile Computing*, vol. 10, no. 6, pp. 821–838, 2011.

O. Tsigkas and F. N. Pavudou, Providing QoS support at the distributed wireless MAC layer: A comprehensive study. *IEEE Wireless Communications*, vol. 15, no. 1, pp. 22–31, 2008.

H. Wang, X. Zhang, F. Nat-Abdesselam and A. Khokhar, Cross-layer optimized MAC to support multihop QoS routing for wireless sensor networks. *IEEE Transactions on Vehicular Technology*, vol. 59, no. 5, pp. 2556–2563, 2010.

Y. Yuan, Z. Yang, Z. He and J. He, An integrated energy aware wireless transmission system for QoS provisioning in wireless sensor network. *Computer Communications*, vol. 29, no. 2, pp. 162–172, 2006.

Chapter 18

Data Aggregation in Wireless Sensor Networks

Introduction

Wireless sensor networks (WSNs) emerge in recent years in many applications but with the drawback of sensor nodes with limited power supply; the issue of energy saving within sensor nodes is always a challenge and issue to be focused upon. Data aggregation and in-network processing are two approaches that are used for saving energy within WSNs, by combining data received from different nodes at some aggregate points en route, and also used for reducing redundancy and minimizing the number of transmission before transmitting the final data to the base station.

Data aggregation in WSNs is a communication means where every node of the network sends their data to the sink node and every intermediate node combines and aggregates all collected or received data with its own packet data by means of suitable aggregation functions as logical, and/or, maximum, or minimum. A routing for an aggregation is a spanning inward arborescence of the communication topology rooted at the sink of the aggregation. Consider the case of time-division multiplexing, where a single node can transmit packet at a time in a particular timeslot. A link schedule of a spanning inward arborescence is the criteria for assignment of timeslot to all the links with two constraints, as, first, a node can transmit only after all its child node finishes its transactions; second, all nodes assigned a common timeslot will transmit data without any interference.

Data aggregation can be performed by means of signal processing and is termed *data fusion*. Data fusion combines some signals to produce an accurate signal. During the process, it also removes signal noise, deploying some techniques. Thus, it can be concluded that data aggregation is used to reduce the required communication level at various levels in order to reduce the total consumption of energy of the network.

The aggregation schedule is thus used for specifying spanning and link schedule for in-arborescence for routing. Latency within aggregation schedule is defined as number of timeslots at which there is occurrence of at least one transmission in that timeslot.

Considering the case of multihop wireless network, aggregation of data is one of the primary communication tasks where a distinguished sink node collects packets from all nodes within the network and combines it with its own packet to aggregate data into single fixed-size packet. One of the major problems in a multihop wireless network is minimum-latency aggregation schedules (MLAs) which is an issue that arises during computing and aggregation schedule with minimum latency within multihop WSNs.

Thus, designing a protocol for data aggregation must focus on eliminating transmission of redundant data to increase the lifetime of energy-constraint WSN. There are cases where sensor nodes sense the same data from environmental variables, which may cause redundancy of data, which needs to be eliminated by means of data aggregation. Thus, data aggregation also removes the redundancy of data.

Taxonomy of Data Aggregation

Data aggregation depends upon various taxonomy:

1. Network lifetime
2. Energy efficiency
3. Data accuracy
4. Latency
5. Data aggregation rate

Network lifetime: Network lifetime is defined as the number of rounds of data aggregation finished till first sensor node is exhausted of its energy. Thus, network lifetime can be defined as the capacity of the first sensor node or group of sensor nodes to run out of energy (battery power) or time (number of rounds) due to failure of one or more sensors.

Energy efficiency: Energy efficiency is defined as the ratio of the amount of data transferred successfully within a sensor network to total energy consumed to transfer those data. In an ideal condition, all sensor nodes should consume the same amount of energy while gathering the data, but in a real situation, sensor nodes consume different energy for transmission of data. Data aggregation works on minimizing the consumption of energy within WSNs.

Data accuracy: The definition of data accuracy differs with the context, which is based on the applications for which the wireless sensors are designed. Data accuracy is defined as the ratio of amount of the data transferred successfully to the total amount of data sent within the network.

Latency: Latency is defined as the time difference between the data produced at the source nodes to the data packets received at the sink nodes. Thus, alternatively, latency can be defined as the time difference between the sending and receiving of data by sensor nodes.

Data aggregation rate: The data aggregation rate is the ratio of the amount of data aggregated successfully to the total amount of data sensed within the network.

History of Data Aggregation

Data aggregation techniques' (DATs') evolution in WSNs has been recorded from the years 2002 to 2018, with a rapid phase, and the research is still ongoing. From the years 2002 to 2015 the focus of study (FoS) of data aggregation varied in following manner: From 2002 to 2003, the prime focus was on network lifetime and energy; in 2004 to 2005 and 2008 to 2009, the focus was on transmission delay, latency, aggregation time, and energy, whereas in 2006 to 2007, the focus was security, bandwidth, and cost; from 2010 to 2018, the focus was on all the points discussed above in addition to other factors like communication redundancy, aggregation rate, error rate and success rate, communication overhead, network stability, delivery ratio, node density, data accuracy, bandwidth, throughput, etc.

In 2002, network lifetime and network-density-based DAT was presented and focused on. In 2003, the focus was on dynamic data aggregation technique, whereas in 2004, the focus was on adaptive data aggregation technique of DAT. In 2005, the focus was on optimal-scheduling-based DAT, which is based on aggregation time and latency. 2006 focused on secure-pattern-based DAT techniques, whereas 2007 focused on sparse data aggregation techniques, concentrating on cost, security, and probability of failure. 2008 adopted the newer DAT technique of linear-distribution-based DAT, which primarily focused on energy, whereas 2009 proposed distributed DAT while focusing on latency and running time. 2010 came with many DAT techniques, like parameter-based, energy-oriented distribution, scalable and dynamic features which focus on cost of communication, energy, latency, etc. 2011 proposed prediction, adaptive clustering, and multi-source temporal DAT, which focuses on redundancy in communication, transmission of packets, error and success rate, etc. 2012 came up with recoverable concealed, two-tier-based clustering approach, tree-based approach for DAT, which primarily focused on energy conservation within the sensor network. 2013 proposed adaptive energy, attribute-aware grouping of cluster-based approach for DAT, focusing on energy, delivery ratio, stability of network, density of nodes, etc. 2014 added to the previous area and proposed shortest-path-based, semantic correlation tree (SCT)-based, adaptive, improved distributed, and latency-based DAT, focusing on transmission overhead, data accuracy and latency, etc. The focus from 2015 onwards has remained the same, with advancements in the previous technologies by focusing on bandwidth efficiency, delay-aware, trust management, clustering, multi-criteria decision-making, learning automata-based DAT, which involves the parameters discussed above with some more features like security, cost of aggregation, energy, etc.

Data Aggregation Techniques

Data aggregation techniques can be of following types:

1. Adaptive data aggregation technique
2. Cluster-based data aggregation technique
3. Concealed-based data aggregation technique
4. Energy-based data aggregation technique
5. Latency-based data aggregation technique
6. Network-lifetime-based data aggregation technique
7. Nature-inspired optimized data aggregation technique
8. QoS-based data aggregation technique
9. Scheduling-based data aggregation technique
10. Tree-based data aggregation technique
11. Network-density-based data aggregation technique
12. Prediction-based data aggregation technique
13. Structure-free data aggregation technique
14. Evolutionary game-based data aggregation technique
15. Hybrid data aggregation technique

Adaptive data aggregation technique: There are many researchers and authors who had proposed different approaches for adaptive data aggregation techniques. There are many protocols like DMAC, AIDA, AEDT, ADANC, DA-MAC which follow different approaches for data aggregation, but the major taxonomy derived from adaptive data aggregation techniques are energy and latency (DMAC), application independence (AIDA), robustness (DA-MAC), reliability (ADA), energy (AEDT), approximation-based (SCT), energy efficiency and data correlation (ADANC).

Cluster-based data aggregation technique: Many researchers had proposed algorithms with different approaches for cluster-based data aggregation. Different protocols proposed under this section are ADA, ADANC, DOC, TTCDA, GCEDA, HDACS, etc. These protocols follow different approaches for data aggregation, but major taxonomy derived from cluster-based data aggregation are adaptive (ADA, ADANC), distributed (DOC), diffusion and clustering (CLUDDA), head-selection scheme–based (ECHSSDA), two-tier clustering (TTCDA), balanced clustering (EEBCDA), group of clusters (GCEDA), bandwidth (BECDA), hierarchical (HDACS), and load distribution.

Concealed-based data aggregation technique: Many researchers had proposed algorithms with different approaches for concealed-based data aggregation. Different protocols proposed under this section are IPHCDA, RCDA, CDAMA, BPDA, DyDAP, etc. These protocols follow different approaches for data aggregation, but major taxonomy derived from concealed-based data aggregation are hierarchical, recoverable, holomorphic public encryption, trust management, balance privacy-preserving, privacy aware.

Energy-based data aggregation technique: Many researchers had proposed algorithms with different approaches for energy-based data aggregation. Different protocols proposed under this section are ESPDA, EPAS, BPDA, holomorphic encryption. These protocols follow different approaches for data aggregation, but major taxonomy derived from energy-based data aggregation are secure pattern, distributed and latency, hierarchical, multi-objective, cooperative, evolutionary game-based, compressed, delivery M2M-based, dynamic, data accuracy, tree-based, structure-free, privacy, remote component-binding, delay-aware, static-clustering, reliability, heterogeneous.

Latency-based data aggregation technique: Many researchers had proposed algorithms with different approaches for latency-based data aggregation. Different protocols proposed under this sections are DMAC, cell-AS. These protocols follow different approaches for latency data aggregation, but major taxonomy derived from adaptive, distributed, conflict aware.

Network-lifetime-based data aggregation technique: Many researchers had proposed algorithms with different approaches for network-lifetime-based data aggregation. Different protocols proposed under this section are MLDA, LAG. These protocols follow different approaches for latency data aggregation, but major taxonomy derived are linear-programming-based, precision-constrained-based, shortest-path-based, multi-criterion-based, learning-automata-based, end-to end delay-based.

Nature-inspired optimized data aggregation technique: Many researchers had proposed algorithms with different approaches for nature-inspired-based optimized data aggregation. Different protocols proposed under this section are ACO-based, DAACA. These protocols follow different approaches for nature-inspired data aggregation, but major taxonomy derived are tree-based, search-space-based, multi-objective, ladder diffusion, network-lifetime, hybrid.

QoS-based data aggregation technique: Many researchers had proposed algorithms with different approaches for QoS-based optimized data aggregation. Different protocols proposed under this section are DAQ, Markov-chain-based, etc. These protocols follow different approaches for nature-inspired data aggregation, but major taxonomy derived are energy, data accuracy, energy, and delay, network-lifetime and fault tolerance, time and energy, cost-based.

Scheduling-based data aggregation technique: Many researchers had proposed algorithms with different approaches for scheduling-based optimized data aggregation. Different protocols proposed under this section are DAS-ST, DAS-UT, DICA, DAS. These protocols follow different approaches for nature-inspired data aggregation, but major taxonomy derived are integrated tree structure, distributed, timeout control scheme, improved distributed, semi-structured and unstructured, delay, and collision free.

Tree-based data aggregation technique: Many researchers had proposed algorithms with different approaches for tree-based optimized data aggregation. Different protocols proposed under this section are ADANC, L-PEADP, AEDT, EEWDA, DST, RAG, etc. These protocols follow different approaches for tree-based data aggregation, but major taxonomy derived are energy, adaptive, multi-objective, collective tree, greedy approach-based, prediction-based, structure-free real-time, ACO-based, evolutionary game-based, scalability-based.

Network-density-based data aggregation technique: Many researchers had proposed algorithms with different approaches for network-density-based optimized data aggregation. Different protocols proposed under this section are ADA etc. These protocols follow different approaches for network density data aggregation, but major taxonomy derived are tree, cluster-based adaptive, network-lifetime-based.

Prediction-based data aggregation technique: Many researchers had proposed algorithms with different approaches for prediction-based optimized data aggregation. Different protocols proposed under this section are MLDA, LAG, RSSI. These protocols follow different approaches for prediction-based data aggregation, but major taxonomy derived are double-queue mechanism, energy-efficient, clustering-based approach.

Structure-free data aggregation technique: Many researchers had proposed algorithms with different approaches for structure-free-based optimized data aggregation. A protocol proposed under this section is RAG. This protocol follows different approaches for structure-free-based data aggregation, but major taxonomy derived are real- time, stochastic time-domain-based, fuzzy-logic-based, attribute-aware-based.

Evolutionary game-based data aggregation technique: Many researchers had proposed algorithms with different approaches for evolutionary game-based optimized data aggregation. Different protocols proposed under this section are EGDAM etc.. These protocols follow different approaches for evolutionary game-based data aggregation, but major taxonomy derived are pixel-level-based, SFB-based approach data aggregation.

Hybrid data aggregation technique: Many researchers had proposed algorithms with different approaches for hybrid-based optimized data aggregation. A protocol proposed under this section is EHDAM. This protocol follows a different approach for hybrid data aggregation, but major taxonomy derived are ant colony optimization-based, clustering-based, energy-aware-based.

Architecture of Data Aggregation

In WSNs, the nodes are selected and distributed in different cluster groups. In order to select the number of nodes within a cluster, parameters like consumption of energy, bandwidth, memory, etc. play a pivotal role. One cluster head is selected among these nodes based on parameters like residual energy etc. This cluster head is responsible for supervising all the nodes within a cluster and also gathering the data from all other nodes within the cluster and sending or transferring this data to other cluster heads for updating. Collection of various data queried from users are checked by data aggregation techniques and by making use of query processor these by are transformed into low-level format. The gathered and aggregated data are stored in the database server. Data cube techniques are used for aggregation of all the data and sending the aggregated data to the base station for future use. This process flow is shown in Figure 18.1.

Categorization of Data Aggregation

The different aggregation techniques presented and adopted by different researchers are as follows:

1. Centralized approach
2. In-network aggregation
3. Tree-based approach
4. Cluster-based approach

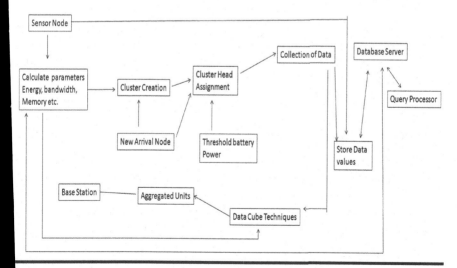

Figure 18.1   Architecture of data gathering and data aggregation in WSNs.

Centralized approach: The centralized approach is centric in nature, where each node sends data to a central node through the shortest path, calculated by various routing protocol in a multihop network. The data is sent to a central node, which is a more powerful node and which aggregates data which can be queried. In this approach, all intermediate nodes send data packets to this centralized node, which transmits a large number of packets for a query in the best case equal to sum of external path lengths for each node.

In-network aggregation: Fasolo et al. (2007) propose an in-network aggregation scheme which uses a global process of gathering and routing information through a multihop network. In this approach, data is processed at intermediate nodes, which reduces consumption of resources and thus increases the lifetime of network. In-network aggregation has two approaches: first, reducing size; second, without reducing size. In reduced-size approach, nodes combine and compress data packets received from its neighboring nodes to reduce the length of packet to be transmitted and forwarded towards the sink node. In the without reducing size approach, the data packets received from different neighboring nodes are merged into a single data packet without processing the value of data.

Tree-based approach: The tree-based approach was proposed by Lee and Wong (2005) and performs the aggregation of data by constructing an aggregated tree, which could be a minimum spanning tree with sink node as root and source node as leaves of the tree. In this tree, each node has a parent node for forwarding the data. Data flow from leaves nodes (from source node) to the parent node (at the sink node) where the aggregation of data takes place by the sink node.

Cluster-based approach: Cluster-based approach is proposed by many researchers as it is a more efficient approach for large networks. In this approach, the entire network is divided into multiple clusters, where each cluster has a cluster head which is selected from other members of the cluster. This cluster head is responsible for the process of aggregation of data received from other members of sensor nodes of cluster and then transmits this aggregated data to the sink node.

Queries in Data Aggregation

In data aggregation, a query is generated at the base station or cluster head and this query is sent towards source nodes, which sense data from environmental variables. These source nodes sense the data required for query to answer and reply with required data. There are three types of queries generated while data aggregation in WSNs: simple query, complex query and event-driven query.

Simple query: These queries use a predictive clause for filtering the sensed data. These queries don't use aggregate functions or aggregators such as max(), min(), sum(), etc. They use simple query for processing of data. Thus, these are non-aggregated queries.

Complex query: Complex queries use aggregate functions and subqueries for processing of data.

Event-driven query: In this type of query, data is returned or reported from source node at regular and periodic intervals of time.

The query processor accepts and generates queries at the client side and retrieves data from the database for specific queries. Thus, query processing extracts information from the database and also optimizes the information simultaneously to provide the information in a fast and efficient manner. The queries extract information from the database in an efficient manner by means of set of commands. The query is based on the database structure, which is known as the database schema, and subsequently filters are used by the query processor. For a huge database, query optimization is done to minimize the resources of systems.

Query optimization is related to relational database management systems (RDBMSs) for transmission of query. Query optimization helps in quick processing of query, minimizing per query cost, increases the performance of process, provides effective procedures for database engine, and also uses less memory. There are two types of query optimization: logical optimization and physical optimization. Logical optimization is used to create a series of relational database whereas physical optimization establishes the functions for effectual query processing.

The processing of queries takes place in three stages: parsing and translation, optimization, and evaluation. Parsing and translation is used to create the query processing engine. This stage is accessed by DBMS to perform the desired task. Query optimization is used to transmit the query to the interior part of the data structure, whereas evaluation is used to create the optimization engine by applicants and perform processing of query.

Conclusion

This chapter focuses on data aggregation in WSNs and various parameters associated with WSNs. This chapter discusses various taxonomy related to data aggregation and discusses the history of data aggregation, including its evolution from 2002 to 2018, considering various factors which were focuses during these periods. The chapter also discusses the architecture of data aggregation in WSNs and categorization in data aggregation which divide it into categories such as centralized approach, in-network aggregation, tree-based approach, cluster-based approach. The chapter focuses on queries and the processing of queries for data aggregation. The chapter also discusses research done by different researchers in the direction of data aggregation and different techniques proposed by researchers to increase the performance of sensor networks.

Related Work

Bista et al. (2009), proposed a data aggregation scheme termed designated path (DP) scheme for WSN, which is also energy efficient in nature. This scheme uses the concept of distribution of work load over the network in order to conserve energy by distributing the workload over the network nodes. The DP scheme determines the set of paths which are run in round-robin fashion such that all nodes participate in the gathering and transferring of the data to the sink node.

Faheem and Boudjit (2010), proposed sensor network multipoint relay SN-MPR mechanism, which is a data aggregation mechanism for mobile sink WSNs. This is a distributed sink location update and a tree-based data-gathering approach deployed for multipoint relay forwarded for location of sink node and queries.

Yang et al. (2011) uses old compression technique and developed a new approach which suppresses the in-network aggregate data and improves the performance of network by means of aggregation.

Zhao and Yang (2010) formalizes the data gathering with multiple mobile collectors (DG-MS) problem by introducing joint design of multiple mobile collectors and spatial-division multiple access (SDMA) techniques.

Zhao et al. (2011) proposed a design of controlled mobility and SDMA technique for the gathering of data using a single SenCar and multiple SenCar and then formalizes them into MDG-SDMA and mobile data gathering with multiple SenCars (MDG-MS) problem.

Nazir and Hasbullah (2010) focused on the hotspot problem and proposed a protocol termed as mobile sink-based routing protocol (MSRP), which is used to increase the lifetime of clustered WSN. This protocol uses the concept of using balanced use of energy within the network and thus increases the lifetime of the network.

Mottola and Picco (2011) proposed a novel routing protocol termed as MUSTER used for many-to-many communication within WSNs. This protocol constructs near optimal routing paths and increases just double the lifetime of WSNs.

Guo and Yang (2012) proposed a new framework termed as data gathering cost minimization (DaGCM), which uploaded the concurrent data constrained by energy consumption, flow conservation, compatibility among sensors, link capacity, and the bound on total sojourn time of the mobile collector at all anchor points.

Mottaghia and Zahabi (2015) propose an algorithm based on working of the low-energy adaptive clustering hierarchy (LEACH) algorithm, which combines the concept of mobile sink and rendezvous nodes to enhance the performance of WSN.

Shiliang Xiao et al. (2015) take consideration of heterogeneous per-node energy constraint to improve the precision in data aggregation by exploiting the tradeoff between data quality and energy consumption to enhance the performance of the network.

Zhao and Yang (2014) utilize mobility for data gathering and joint energy replenishment. They employed SenCar, which is a multi-functional mobile entity which collects data while roaming over the field using short-range communication, and it is also used as an energy transporter which charges static sensors on its migration tour using wireless energy transmission.

Soltani et al. (2014) proposed an approach of data fusion which efficiently uses resources for large WSNs. Data fusion determines reduced node set to be active in the network, which reduces consumption of network resources.

Gong and Yang (2014) construct a reliable model which is a tree-based approach for the gathering of data which schedules transmission for the links on the tree and assigns transmission power to each link accordingly.

Yuan et al. (2014) has proposed a clustering method for data aggregation in WSNs which is based on data density correlation degree.

Xiang et al. (2013) has proposed a compressed data aggregation scheme which uses a compressed sensing (CS) technique in order to achieve higher energy efficiency and recovery fidelity within WSNs with arbitrary topology.

Guo et al. (2013) proposed a framework which works jointly for replenishment of wireless energy and anchor point-based mobile data gathering in WSNs by view of various sources of energy consumption and time-varying nature of energy replenishment.

Mathapati et al. (2012) developed a new energy- efficient routing protocol termed as energy efficient reliable routing protocol (EERRP) which uses a data aggregation technique for WSNs. Data aggregation is used for the collection and aggregation of data in an efficient manner in order to increase the lifetime of network.

Parameters Considered for Designing Data Aggregation Techniques

Energy efficiency: Due to the heavy density of nodes within the sensor network, there are chances of data redundancy due to spatial correlation which may consume energy and lead to more consumption of energy, bandwidth, and time by the sensor network. Data fusion can be used to alleviate this problem. The data aggregation process aggregates the data and thus reduces redundancy and sends the aggregated data to the base station for processing. Thus, for designing a data aggregation technique, efficiency of energy is one of the most important parameters that needs to be taken care of.

Time-bound: Time constraint is one of the most important factors while designing for data aggregation techniques for WSNs. Time constraint here refers to the time involved in the aggregation of data and the reporting of aggregated data to the base station. So, while designing the aggregated algorithms, time constraint for aggregation of data and sending data to the base station should be kept as low as possible in order to increase the efficiency of the network.

Application-oriented: Data aggregation is applied based on the requirement of the application. There are certain applications which require high accuracy without having more concern about delay or time constraint, whereas there are many applications which have more stress on time constraints rather than on accuracy. Hence, while designing a data aggregation protocol, the requirements of applications should be kept in mind and should be taken care of.

QoS support: With the advancement in usage of WSN technology, the need for support for QoS increases accordingly. QoS parameters are used for improvement in delay, data accuracy, bandwidth, throughput, etc. QoS parameters also play a vital role while dealing with multimedia type of applications in WSNs.

Resource utilization: For increasing the lifetime of the sensor network in data aggregation, it is required that the resources available with WSNs need to be utilized efficiently. So, algorithm designed should be such that it should use the maximum amount of resources available with the network.

Load balancing: In case of a heavy traffic load, traffic can be split into multiple paths which are discovered during the route discovery phase using data dissemination algorithms. But concurrent use of multiple alternate paths may result in degradation of performance of the network due to inter-path interference. Use of multipath can use specific data aggregation or data fusion process such that the data can be merged together before sending to the sink node for efficient usage of bandwidth, energy, etc.

Reliability and fault tolerance: WSNs may be time varied in nature and may also have dynamic topology. So, for such networks, reliability and fault tolerance are one of the major challenges that need to be taken care of before designing algorithms for data aggregation.

References

R. Bista, Y.-K. Kim and J.-W. Chang, A new approach for energy-balanced data aggregation in wireless sensor networks, In: *IEEE Ninth International Conference on Computer and Information Technology*, China, Volume 2, 2009.

Y. Faheem and S. Boudjit, SN-MPR: A multi- point relay based routing protocol for wireless sensor networks, In: *IEEE/ACM International Conference on Green Computing and Communications & IEEE/ACM International Conference on Cyber, Physical and Social Computing*, 2010, pp. 761–767.

E. Fasolo, M. Rossi, J. Widmer and M. Zorzi, In-network aggregation techniques for wireless sensor networks: A survey, *IEEE Wireless Communication*, vol. 14, no. 2, pp. 70–87, 2007.

D. Gong and Y. Yang, Low-latency SINR-based data gathering in wireless sensor networks, *IEEE Transactions on Wireless Communications*, vol. 13, no. 6, pp. 3207–3221, 2014.

S. Guo, C. Wang and Y. Yang, Mobile data gathering with wireless energy replenishment in rechargeable sensor networks, In: *Proceedings IEEE INFOCOM*, 2013, pp. 1932–1940.

S. Guo and Y. Yang, A distributed optimal framework for mobile data gathering with concurrent data uploading in wireless sensor networks, In: *Proceedings IEEE INFOCOM*, 2012, pp. 1305–1313.

M. Lee and V. W. S. Wong, An energy-aware spanning tree algorithm for data aggregation in wireless sensor networks, In: *IEEE PacRrim*, 2005.

B. S. Mathapati, S. R. Patil and V. D. Mytri, Energy efficient reliable data aggregation technique for wireless sensor networks, In: *International Conference on Computing Sciences*, 2012, pp. 153–158.

S. Mottaghia and M. R. Zahabi, Optimizing LEACH clustering algorithm with mobile sink and rendezvous nodes, *International Journal of Electronics and Communications*, vol. 69, no. 2, pp. 507–514, 2015.

L. Mottola and G. P. Picco, MUSTER: Adaptive energy aware multisink routing in wireless sensor networks, *IEEE Transactions on Mobile Computing*, vol. 10, no. 12, pp. 1694–1709, 2011.

B. Nazir and H. Hasbullah, Mobile sink based routing protocol (MSRP) for prolonging network lifetime in clustered wireless sensor network, In: *International Conference on Computer Applications and Industrial Electronics (ICCAIE)*, 2010.

M. R. Soltani, M. Hempel and H. Sharif, Data fusion utilization for optimizing large-scale wireless sensor networks, IN: *IEEE ICC 2014 – Ad-hoc and Sensor Networking Symposium*, 2014, pp. 367–372.

L. Xiang, J. Luo and C. Rosenberg, Compressed data aggregation: Energy-efficient and high-fidelity data collection, *IEEE/ACM Transactions on Networking*, vol. 21, no. 6, pp. 1722–1735, 2013.

S. Xiao, B. Li and X. Yuan, Maximizing precision for energy-efficient data aggregation in wireless sensor networks with lossy links, *Ad Hoc Networks*, vol. 26, pp. 103–113, 2015.

C. Yang, Z. Yang, K. Ren and C. Liu, Transmission reduction based on order compression of compound aggregate data over wireless sensor networks, In: *6th International Conference on Pervasive Computing and Applications (ICPCA)*, 2011, pp. 335–342.

F. Yuan, Y. Zhan and Y. Wang, Data density correlation degree clustering method for data aggregation in WSN, *IEEE Sensors Journal*, vol. 14, no. 4, pp. 1089–1098, 2014.

M. Zhao, J. Li and Y. Yang, A framework of joint mobile energy replenishment and data gathering in wireless rechargeable sensor networks, *IEEE Transactions on Mobile Computing*, vol. 13, no. 12, pp. 2689–2705, 2014.

M. Zhao, M. Ma and Y. Yang, Efficient data gathering with mobile collectors and space-division multiple access technique in wireless sensor networks, *IEEE Transactions on Computers*, vol. 60, no. 3, pp. 400–417, 2011.

M. Zhao and Y. Yang, Data gathering in wireless sensor networks with multiple mobile collectors and SDMA technique sensor networks, In: *WCNC Proceedings*, 2010, pp. 1–6.

INTERMITTENTLY CONNECTED DELAY-TOLERANT WIRELESS SENSOR NETWORKS (ICDT-WSNs)

IX

Chapter 19

Data Collection in Sparse Wireless Sensor Networks with Mobile Nodes

Due to continuing technological advancements, wireless sensor networks (WSNs) have been widely accepted as a popular platform for a variety of applications ranging from monitoring to tracking, viz. numerous applications like habitat monitoring, pollution monitoring and control, structural monitoring, health monitoring, precision agriculture, surveillance, etc. In addition to the above applications, WSNs also play a very significant role in applications like disaster management and recovery, intrusion detection, etc. Nodes in the WSN are meant to collect the samples from their surroundings, perform some sort of simple processing over the data collected, and then finally transport the processed information to a centralized processing system usually termed as the sink or base station via direct or multihop data forwarding depending on the employed routing technique.

In normal application scenarios, wireless sensor networks tend to contain a huge number of static nodes which are deployed densely, i.e., close to one another, hence facilitating multihop communication among them. However, due to the increasing acceptability of WSNs, applications have been identified which require sensor nodes to be deployed sparsely with few or more (depending upon the nature of the applications) being mobile. In this new network scenario, some nodes are mobile and their responsibility is to collect data from other nodes and convey them to the base station. For example, in the animal monitoring system, where nodes are installed over wild/domestic and roaming animal to collect environmental and biological data, or in the infantry networks, in which nodes are carried by military personnel for security and monitoring reasons, contrary to the traditional WSNs, networks show a number of different characteristics viz. sparseness, mobility, extreme energy constraint, and huge delay incurred in data delivery. Here, sparseness refers to the number of nodes deployed; mobility refers to the ability of nodes to move from one place to another; extreme energy constraints refers to the fact that the batteries of the nodes cannot be replaced easily and they are required to remain functional over a very long period of time; and the delay refers to the event that data delivery may take a long time and, therefore, networks must consider this fact.

To meet the needs of such applications, mobility has been introduced to WSNs which satisfies the requirements of the concerned application by providing improved connectivity, reduced cost, increased reliability, and enhanced energy efficiency as follows.

Connectivity: Due to the lower number of nodes and their limited transmission range, connectivity among the nodes in the WSN may become an issue; but, when featured with mobility, mobile elements (MEs) cope with the distant unattended regions.

Cost: As the number of nodes in this variant of WSN is quite low, the cost of the network reduces. Mobility might increase the cost, but through the use of existing mobile elements, e.g., buses, cars, trackers, travelers etc., the network cost might further be reduced.

Reliability: The dense population of nodes in traditional networks may result in contention among the nodes and interferences as well due to the close nodes' proximity. Also, due to the usual multihop communications approach, messages may suffer with data loss in every hop towards the destination. When featured with mobility, mobile elements visit the nodes in order to collect the data using single-hop communication, which prevents not only contention and interferences but also incremental data losses.

Energy efficiency: As the mobile nodes visit each isolated region of the network in order to collect the region-specific data, nodes in the network directly communicate to these mobile elements instead of communicating to the sink; hence, the communication cost is reduced and energy efficiency is achieved.

However, nothing comes at free and there is always a tradeoff. Mobility introduces a number of challenges which are not present in the traditional sensor networks, as follows.

Detection of message forwarding opportunity: Since the nodes can transfer their data only when they are visited by the mobile elements, accurate detection of their presence is mandatory and the most required, especially when the duration of visit is quite short. Such opportunity of data transfer is termed as contact in the taxonomy.

Power management: Keeping nodes awake all the time may drain the their battery at a higher rate; hence, they are awoken only when contact is anticipated. Knowledge of mobility pattern here plays a very crucial role.

Reliability: As the contact duration may be of very short span, all the information must be transferred correctly.

This chapter first provides a detailed overview of the changed network scenario as indicated above, i.e., wireless sensor networks with mobile nodes; then, a network taxonomy is provided, followed by a detailed classification of existing data collection processes in the subsequent sections.

Overview of WSNs with Mobile Elements

Wireless sensor networks with mobile elements, popularly known as WSN-MEs is composed of three main components—regular sensor nodes, sink, and the intermediary nodes, with any of these (but at least one) being mobile as shown in Figure 19.1.

Regular Nodes

These nodes are the normal nodes whose only responsibility is to sample and process the surrounding environment in order to forward them to the sink for further processing for end-user access. They may or may not be mobile; as in Figure 19.1, regular nodes are static in nature.

Sink

As in the traditional wireless sensor network, the sink is final destination of the information sent from the network nodes. Here, the sink may be either static or dynamic. When the sink is found to be dynamic in the network, it visits the nodes itself in order to collect the data, and if it is static, data from the nodes are delivered by the intermediary nodes, as explained later.

Intermediary Nodes

Intermediary nodes, popularly known as mobile mules or support nodes, are neither the source nor the destination of the message but play very crucial roles of either data collector or mobile gateway. These nodes are meant to support either the network operation or the data collection. In Figure 19.1, the mobile mules are assisting in the data collection by visiting different regions in the network where sensors are deployed sparsely.

As stated above, mobility may be implemented at any of these three levels, but implementation on at least one level is mandatory, as in the absence of mobility, the network is reduced to traditional wireless sensor network. Moreover, implementing mobility provisions fewer numbers of nodes as compared to the huge number of

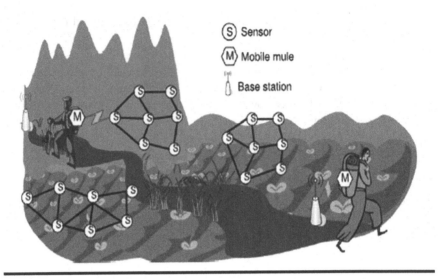

Figure 19.1   An instance of WSN-MEs system. (From Tseng et al., 2013.)

nodes densely deployed in the network. Mobile elements take care of the message transportation to the sink.

Mobile Elements

Depending upon the level of mobility implementation, there can be different types of Mobile elements in the system, as follows (Francesco et al., 2011).

Relocatable Nodes

In the WSN-MEs system, the regular nodes with the ability to change their location with an objective either to improve the coverage or to strengthen the connectivity are known as relocatable nodes, as shown in Figure 19.2. Such nodes are not meant to carry the data like mobile mules as explained above, but the objective is to bring changes in network topology in order to avoid coverage holes and to improve communication reliability as well as energy efficiency. The relocatable nodes become static and start reacting like regular sensor nodes once they are shifted to some new location. They never actively participate in the data collection process but only provide a mobility-assisted approach to the network by filling the gaps.

Mobile Sinks

When the mobility is implemented over the sink, which is the final destination of messages generated by regular nodes, the sink is termed as the mobile sink. The

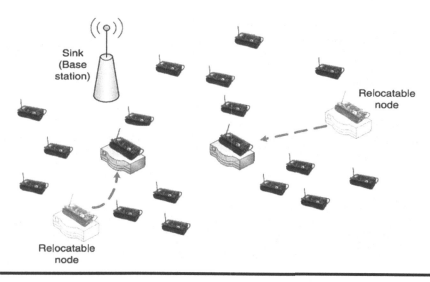

Figure 19.2   WSN-MEs with relocatable nodes. (From Francesco et al., 2011.)

data collected by the mobile sinks are either accessed by the end user via wireless connections for some application-specific decision-making process or those are consumed by the sink itself for some preconfigured application-specific processing. An instance of mobile sinks is shown in Figure 19.3. Here, the rendezvous points are optional entities which are meant to assist mobile sinks in collecting the data from the concerned regions. Rendezvous points are installed on the path of mobile sinks to ease the data collection process as they collect data regularly from the

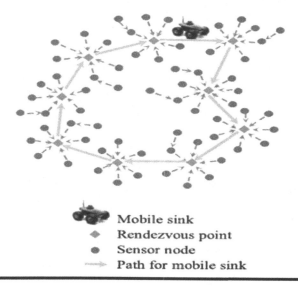

       Mobile sink
◆   Rendezvous point
●   Sensor node
⟶   Path for mobile sink

Figure 19.3   WSN-MEs with mobile sinks. (From Kaswan et al., 2017.)

nearby surrounding sensors and hand over the collected data as soon as they come into the contact with mobile sinks.

Mobile Relays

Mobile relays are the intermediary nodes which are also known as mobile mules. These nodes are responsible for collecting the data from the regular nodes in view of delivering them to the sink. A relay node acts as just a message forwarding node and not the endpoint of communication. As soon as the mobile relay comes into the contact with the sensor nodes, the nodes transfer their data to the mobile relay/mule which in turn moves towards the base station and transfers the collected data when it comes into contact with the sink, as depicted in Figures 19.1 and Figure 19.4.

Sometimes, both mobile relays and mobile sinks are collectively referred to as mobile data collectors (MDCs) in the literature, but, due to their completely different roles, here they are defined separately.

Mobile Peers

In networks where mobility is implemented in the regular nodes who are responsible for sampling the environment, nodes are termed *mobile peer nodes*. Here, mobile peers are the regular nodes who perform both the actions: message generation and message relay to the sink node. However, the sink may also be mobile. As soon as a mobile peer comes into with contact of sink, it not only transfers its own data but also the data from other peer nodes collected while being in contact with them. An example of such WSN-MEs with mobile peers is shown in Figure 19.5, which is an instance of an animal monitoring system.

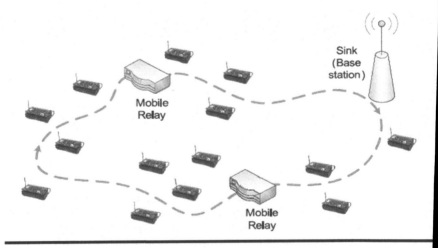

Figure 19.4 WSN-MEs with mobile relays. (From Francesco et al., 2011.)

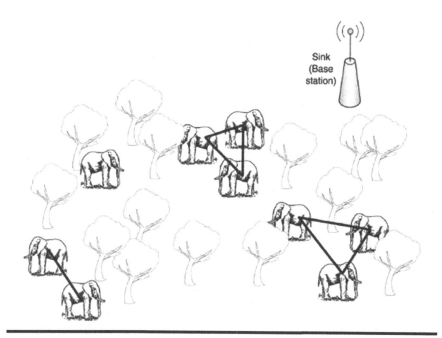

Figure 19.5   WSN-MEs with mobile peers—An application instance.

Once the data is communicated to the sink, it gets removed from the nodes' memory in an attempt to save memory for further operations.

The Process of Data Collection in WSN-MEs

The process of data collection in the WSN-MEs system can be summarized in Figure 19.6.

The detailed taxonomy which is required for imbibing the concept summarized in Figure 19.6 is given below:

Contact: A sensor node is said to be in *contact* with some other node when they both can exchange message in between.

Contact area: Contact area refers to the region in which nodes are said to be contact with one another.

Discovery: As a matter of fact, nodes are required to detect the presence of mobile elements before they can forward their data. The process of finding or confirming the presence of forwarding opportunities via locating the mobile elements is referred to as *discovery*.

Residual contact time: Mobile elements come into contact with regular nodes for a much shorter period of time, and the process of discovery also takes some time. So, the effective period for forwarding messages can be computed as the difference of total contact duration and time taken in the discovery process. This total effective duration for the data transfer from regular nodes to mobile elements is known as *residual contact time*.

Data transfer: Data transfer refers to the process of message exchange between the nodes in which at least one node is mobile.

Routing: Routing refers to the activities viz. appropriate path selection etc. in a process of message forwarding towards the mobile elements.

Now, the data collection process in the WSN-MEs system can be summarized in the following paragraph.

The regular sensor nodes wait for the forwarding opportunities to arise, and as soon as a mobile element comes into contact with the nodes, after confirming their presence (through the process of discovery) into respective contact areas, the nodes start the data transfer process.

In addition to the above discussed concepts, it can be easily intuited that the mobility of mobile elements can have a great impact over the entire data collection process. The most significant aspect of the mobility which may even drive the entire process accordingly is the feature of mobility being controllable or uncontrollable, i.e., controllability.

In the case of controlled mobility, mobility can be exploited to facilitate the process with fewer challenges. For example, since the trajectory and speed are already known and can further be controlled, the discovery process may be easily scheduled in order to lengthen the contact duration and to improve the energy efficiency. In this way, movement can be effectively treated as an additional factor to ease the data collection process.

With the uncontrolled mobility, prior knowledge of mobility patterns can be used to guide the data collection process efficiently. There are basically two mobility patterns under this category of uncontrolled mobility: *random* and *deterministic*. Here, random mobility pattern refers to the event that MEs are visiting the nodes without

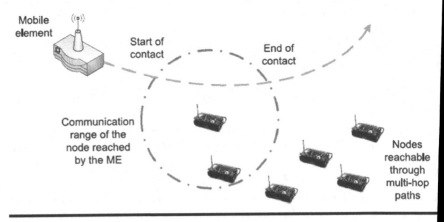

Figure 19.6   Data collection in WSN-MEs. (From Francesco et al., 2011.)

showing any definite regularity, but probabilistically; whereas, deterministic mobility patterns indicate that some sort of regularity has been demonstrated in visiting the nodes by MEs. Moreover, through some prior knowledge of movement patterns, nodes can be configured to schedule themselves for discovery process when there are high chances of coming into contact with MEs instead of waiting all the time.

Thus, from the above discussion, the entire process of data collection can be seen as being comprised of the four phases, namely, *discovery, data transfer, routing,* and *mobility control,* with each being discussed in detail below.

Discovery

As indicated above, discovery refers to the detection of message forwarding opportunity to the nodes. A node can only transfer its data to the MEs (the collecting nodes) if they find themselves within their contact, and that's why in addition to accurate discovery, timely discovery is also required in order to utilize the most of contact duration. There are basically two different approaches for the discovery process: mobility independent and mobility aware.

As implied by the name, mobility independent discovery methods do not require information on the mobility pattern, whereas this is required in the mobility aware discovery (Francesco et al., 2011).

Mobility-Independent Discovery Schemes

The mobility independent discovery schemes can further be classified as follows:

1. Scheduled rendezvous discovery
2. On-demand discovery
3. Asynchronous discovery

Scheduled Rendezvous Discovery

Scheduled rendezvous discovery refers to a strong consensus between the nodes and mobile elements about the contact timing. This model can be implemented when the mobile elements are found to follow a strict schedule and the nodes are aware of the same, e.g., in the applications where collecting nodes are installed over public transportation following strict schedule of visiting sensor nodes deployed in the sensing field (Chakrabarti et al., 2003).

On-Demand Discovery

On-demand discovery refers to awakening the nodes in the network through a process initiated by mobile elements. The two approaches widely used in this category are the activation through low-power radio and radio-triggered activation.

In the first method, nodes are directed to regularly sense the MEs' presence on low-power radio channels, called the paging channels, whereas in the second method, high-energy wake-up messages are sent by the MEs in order to get the sensors activated.

Asynchronous Discovery

Without any need of consensus among the communicating nodes, as in the previously mentioned scheduled rendezvous discovery, asynchronous discovery requires MEs to broadcast their presence periodically and, if heard by the nodes, the nodes start the data transfer. Moreover, nodes also follow their sleep/wake-up schedule, and if they do not hear any ME broadcast, they go into sleep mode again.

Mobility-Aware Discovery

In mobility-aware discovery of mobile elements, knowledge of mobility patterns is utilized to improve the efficiency of the discovery process. When the pattern is exploited in the discovery process, nodes can be scheduled accordingly to be active when encounters with MEs are most likely; they can be put in the sleeping mode for rest of the time. However, nodes are directed for the training phase first, where they learn the MEs' mobility pattern by collecting their arrival data to be exploited later in deciding their wake/sleep schedule.

Data Transfer

Data transfer refers to the process of message exchange between the nodes and mobile elements once their (MEs') presence is confirmed through the discovery phase. Data transfer involves a single-hop communication between the node and mobile element. Moreover, since data transfer involves mobility as its inherent component, adequate care must be taken while implementing this, i.e., the impact of the mobility must be kept under consideration while dealing with data transfer in WSN-MEs. Along with the mobility, there arises issues viz. speed and varying distance between the communicating parties which, if not taken seriously, might degrade the overall network performance.

Routing

Routing refers to the activities viz. appropriate path selection etc. in a process of message forwarding towards the mobile elements. The problem of routing arises when the network is comparatively dense or the network is partitioned into isolated sparse sub-networks and accordingly the full multihop or partial multihop can be implemented. There are two major classes of routing techniques in the context of WSN-MEs (Francesco et al., 2011):

1. Flat routing
2. Proxy-based routing

However, both the classes require the routes to be computed and updated adaptively while traversing through the network.

Flat Routing

In flat routing, every node in the network is of peer authority and behaves in the same way, i.e., there is no such extra and special role to be assigned to any of the nodes in network. Many routing schemes have already been introduced by the researcher community either by modifying the existing WSNs' solutions into the context of WSN-MEs or by devising novel solutions to cope with the changed network scenario as compared to the traditional WSNs as follows.

> *MobiRoute*: MobiRoute (Luo et al., 2006) is a modification of the MintRoute protocol (Woo et al., 2003). In the MintRoute protocol, a tree structure is developed for the static sensor nodes for data collection, with the path being evaluated using a metric attempting successive minimization of (re)transmission. Moreover, MintRoute continuously monitors the link quality and computes and updates the routing tree. When compared to MintRoute, MobiRoute utilizes three different mechanisms to cope with the mobility of mobile elements. First, in order to monitor the link health severely affected by the moving MEs, it employs the beacon messages and timeouts. Then, it restricts the tree reconstruction phase, hence reducing the overhead, and survives with the temporary suboptimal routing paths. Lastly, bandwidth throttling and data buffering are implemented to minimize the data losses due to moving MEs.
>
> *WEDAS*: WEDAS, i.e., the Weighted Entropy DAta diSsemination protocol, has been introduced in the context of WSN-MEs (Ammari and Das, 2005) and utilizes an information–theoretic approach. The scheme selects message forwarders between the nodes and mobile elements based on two different parameters: residual energies of the nodes and the position of MEs.

Proxy-Based Routing

In proxy-based routing, some of the nodes from the network are selected as proxies or gateways whose role is to represent a group of nodes, i.e., they act like communication bridges between the regular nodes and the mobile elements in order to facilitate data transfer. The specially appointed nodes, termed as proxies or gateways or anchors or rendezvous points, are meant to collect data from the representing portion of network, and as soon as they come into contact with MEs, data transfer starts. The scheme works well in both variants of WSN-MEs, viz. where nodes are deployed comparatively dense, i.e., dense WSN-MEs, and also where the network can be partitioned into isolated sub-networks, i.e., sparse WSN-MEs. For example, two-tier data dissemination (TTDD) has been developed in Luo et al. (2005) for dense WSN-MEs and scalable energy-efficient asynchronous dissemination (SEAD) protocol was developed in Kim et al. (2003) to cope with the sparse WSN-MEs.

Mobility Control

As stated earlier, mobility of the mobile elements can have a great impact over the entire data collection process and can even drive the entire data collection process in a very significant way. Mobility can either be controllable or uncontrollable. When a system adapts uncontrolled mobility, there remains no other way but to tune up to the way ME moves. But, with the controllable mobility, a more flexible data collection process can be designed. Controlling MEs' mobility simply refers to defining their trajectory, i.e., the path followed, or defining the speed of the mobile elements.

References

H. Ammari and S. Das, Data dissemination to mobile sinks in wireless sensor networks: An information theoretic approach. In: *Proceedings of the 2nd IEEE International Conference on Mobile Ad Hoc and Sensor Systems*, 2005, pp. 8–314.

A. Chakrabarti, A. Sabharwal and B. Aazhang, Using predictable observer mobility for power efficient design of sensor networks. In: *Proceedings of the 2nd International Workshop on Information Processing in Sensor Networks*, 2003, pp. 129–145.

M. Di Francesco, S. K. Das and G. Anastasi, Data collection in wireless sensor networks with mobile elements: A survey. *ACM Transactions on Sensor Networks*, vol. 8, no. 1, pp. 7, 2011.

A. Kaswan, K. Nitesh and P. K. Jana, Energy efficient path selection for mobile sink and data gathering in wireless sensor networks. *International Journal of Electronics*, vol. 73, pp. 110–118, 2017.

H. S. Kim, T. F. Abdelzaher and W. H. Kwon, Minimum-energy asynchronous dissemination to mobile sinks in wireless sensor networks. In: *Proceedings of the 1st ACM Conference on Embedded Networked Sensor Systems*, 2003, pp. 193–204.

H. Luo, F. Ye, J. Cheng, S. Lu and L. Zhang, TTDD: Two-tier data dissemination in large-scale wireless sensor networks. *Wireless Networks*, vol. 11, no. 1–2, pp. 161–175, 2005.

J. Luo, J. Panchard, M. Piorkowski, M. Grossglauser and J. Hubaux, MobiRoute: Routing towards a mobile sink for improving lifetime in sensor networks. In: *Proceedings of the 2nd IEEE International Conference on Distributed Computing in Sensor Systems*, 2006, pp. 480–497.

Y.-C. Tseng, F.-J. Wu and W.-T. Lai, Opportunistic data collection for disconnected wireless sensor networks by mobile mules. *Ad Hoc Networks*, vol. 11, pp. 1150–1164, 2013.

A. Woo, T. Tong and D. Culler, Taming the underlying challenges of reliable multihop routing in sensor networks. In: *Proceedings of the 1st ACM Conference on Embedded Networked Sensor Systems*, pp. 14–27, 2003.

Chapter 20

Opportunistic Networks in Disaster Management

One of the major requirements in mobile ad hoc networks (MANETs) is the establishment of an end-to-end connection between the communicating nodes before exchange of any message. However, this may not always be the case, especially in scenarios with poor network links, mobile nodes, asynchronous data transfer, and high error rates, i.e., in the challenged networks. More illustratively, it might be very possible that the network is undergoing frequent topological changes either due to high node mobility or high rate of nodes' failure; expecting an end-to-end connectivity between the hosts in such a network scenario would be meaningless. Similarly, intermittent connectivity breaches due to poor communication links cause an end-to-end connectivity between the hosts to not always exist. Also, asynchronous data transfer and high error rates may result in long end-to-end delay.

For example, if we consider the networks of mobile devices being carried by people, more specifically called pocket-switched networks (PSNs), characterized by the moving nodes (as per the human mobility pattern), with high processing and storage capacity but with limited energy resources, instantaneous end-to-end connections are required for the exchange of any message which might not be available always, and the packet exchanges get deferred by long delays.

Another class of such networks is the vehicular ad hoc networks (VANETs), which are characterized by nodes moving with high speed and communicating with other high-speed moving nodes and other roadside units (RSUs) to explore better route decisions. Since the nodes can communicate only when they come into close contact with one another, communication is said to be opportunistic and may suffer with huge delays.

Lastly, when we think of interplanetary networks (IPNs), i.e., the networks of satellites moving around the planets, these networks also exhibit the suffering of huge delays due to long distances between the objects and intermittent connectivity due to the movement and rotation of planets and satellites.

These networks, which are characterized by the sparse density of nodes, unpredictable nodes' movements, poor network link, and long period of disconnections are called opportunistic networks (OppNets). The notion of OppNets came with the idea of IPNs where the delay-tolerant networks (DTNs) was born (Burleigh et al., 2003). DTNs are developed as overlay on the regional networks (i.e., the networks with homogeneous communication characteristics) and their main aim is to provide interoperability among these regional networks. They operate over the TCP/IP protocol stack and serve as a gateway for the participating networks over the links constrained by delay or disruption. Opportunistic networks cover a broader perspective, as they not only treat the disconnections and disruptions among networks but also support the nodes. Moreover, they do not mandate the use of TCP/IP (Mota et al., 2014). In the OppNets, the messages to be transmitted are provisioned to be stored in the intermediary nodes, which preserve the message until they get a proper opportunity to transmit them. Once the message-carrying nodes come into close/direct communication with some other node, messages based on priority are transferred to the visiting nodes. The process of keeping the message until a suitable contact is made with some eligible node is called custody, and the entire paradigm is referred to as the *store-carry-forward* paradigm.

From the above discussion, it can be easily intuited that the most appropriate use of opportunistic networks could be in emergency scenarios where quick and coordinated responses must be provided to the field rescue team to improve the efficiency of the system in order to save as many lives as possible. The very nature of OppNets, like being infrastructure-less and delay tolerant, makes them suitable for their use in emergency situations.

In the upcoming sections of this chapter, first, a detailed discussion is made on the types of opportunistic networks, and then a classification of OppNet-based applications is done with an emphasis on disaster management. At last, a taxonomy of existing forwarding protocols in opportunistic networks is discussed.

Types of Opportunistic Networks

Unlike traditional networks, opportunistic networks are more prone to frequent link failures and long periods of network disconnection; hence, the traditional approach of assuming an always-on connectivity between the hosts is not suitable for such networks. Consequently, *store-carry-forward*, as discussed above, is followed in OppNets. In the store-carry-forward paradigm, data retrieved by the nodes are kept with themselves until an appropriate opportunity arises in terms of *contact* between them. Contact is basically a term which signifies the establishment of an instantaneous link between the communicating nodes. Once the link is established, data is forwarded immediately to the counter party; this is referred to as *custody transfer* in the literature.

As per the applications found in the literature, opportunistic networks can be classified as follows:

1. Delay-tolerant networks
2. Intermittently connected networks or challenged networks

Delay-Tolerant Networks

DTNs are developed as overlay on the regional networks (i.e., the networks with homogeneous communication characteristics) whose main aim is to provide interoperability among these regional networks (Fall, 2003). In the discussion led by Delay-Tolerant Networking Research Group (DTNRG) in 2012, an additional layer entitled the bundle layer was introduced between the application layer and the transport layer, not only to provide transparency among a number of regional networks but also to mask the disconnection and delay from the application layer. The bundle layer encapsulates the application data into bundles and forwards them to the transport layer. To compensate for the network interruptions, DTNs utilize the store-carry-forward paradigm, in which the message is transferred to the next node only when an appropriate link is established between them. Such transfer of packets is termed as custody transfer in the literature and ensures end-to-end reliability and security in the network. Custody transfers prevents loss of data and corruption through time-to-acknowledge (TTA) retransmission and time-to-live (TTL) facility. More illustratively, the node willing to transfer the custody of the data held initiates the process by starting the TTA retransmission timer. If the counter node successfully receives and accepts the custody, it reverts back to the transmitting node with an acknowledgment signifying the success of custody transfer. However, if the TTA timer expires and the respective acknowledgment is not received by the sender, custody transfer is declared to have failed and it is reattempted by the sender. Other than the TTA retransmission timer, the TTL timer also plays an important role in deciding how long the bundle is to be kept in the carrying node. The node supporting the custody transfer maintains the data till the expiration of the time-to-live timer.

Moreover, it can be easily intuited from the above discussion that the primary requirement to support custody transfer is the ability to be equipped with enough storage so that the bundles can be accommodated well inside.

Delay-tolerant networks contain basically two types of networks: underwater networks and interplanetary networks.

Underwater networks: In the underwater network, a variable number of nodes are deployed sparsely in the water to achieve objectives, viz. oceanographic data collection, pollution monitoring, offshore exploration, disaster forecasting and prevention, and tactical surveillance applications (Akyildiz et al., 2005). Moreover, contrary to the terrestrial networks, acoustic waves are used for communication in such networks, as the radio frequency (RF) communication is more prone to attenuation in aquatic environments. Underwater networks suffer with longer propagation delay (as the nodes move along the tides and currents), with limited and non-replaceable battery (as in the terrestrial networks), with the high sparsity of the nodes, and with the high link-error rates.

Interplanetary Networks: As dictated above, interplanetary networks are deployed to facilitate communication among satellites and the earth or even other planets. They consist of nodes deployed on satellites or on other space objects to achieve the above stated objective. As a matter of fact, IPN suffers with a very long and variable propagation delay, as the distance between two communicating hosts may be very large and also the two separate types of movement (rotation and circular displacement) of objects over which nodes are mounted results in intermittent connectivity among the nodes. However, easy predictability of next contact to occur differentiate the interplanetary networks from other opportunistic networks, where such computations can never be done a priori.

Intermittently Connected Networks or Challenged Networks

Networks violating one or more assumptions of basic internet architecture, viz. always-on end-to-end connectivity between the hosts or bounded roundtrip time etc., are categorized as challenged networks (Fall, 2003). Such challenged networks may have one or more following characteristics:

- High end-to-end path latency
- Frequent end-to-end disconnections
- Limited resources
- Limited life expectancy

High node mobility and node sparsity result in the first two characteristics cited above, whereas the latter two are the results of constraints imposed by their size and environment in which the network is to be deployed. Hence, factors like high node mobility, high node sparsity, link unreliability, etc. restrict the use of solutions developed for traditional MANETs in challenged networks; this motivates researchers to search for suitable solutions with regard to this new network scenario.

Here, three variants of challenged networks, namely, VANETs, PSNs, and mobile wireless sensor networks (MWSNs) are discussed in detail.

Vehicular ad hoc networks: VANETs are comprised of communicating nodes mounted on highly movable terrestrial verticals. The nodes exchange packets with one another when and where required and also communicate to the RSU in view of providing sound and safe road transportation. Moreover, VANETs can be considered as a variant of traditional MANETs in which characteristics of the MANETs are improved to a greater extent, e.g., in comparison to MANETs, mobility is very high in VANETs, but predictable in nature. Similarly, the network topology is highly variable in time and space when compared to the traditional MANET as the current traffic conditions like traffic jams or slowly moving traffic affects the network topology a lot. Also, since the nodes' communication ranges are limited, the network may get partitioned. However, energy is never a constraint in VANETs when compared to the other variants of challenged networks.

Pocket-switched networks: Pocket-switched network are a variant of traditional MANETs in which portable devices like mobile phones and personal digital assistants (PDAs) are carried by humans connected via Bluetooth or Wi-Fi, thus creating a sparse and intermittent network as the regular movements of the human restricts a dedicated line to be created among the host. Other than communicating with one another, these nodes can also communicate with the fixed nodes providing services such as internet connectivity etc. PSNs enable two nodes to communicate opportunistically only, i.e., nodes can exchange their data only when they come into close contact with each other.

Mobile Wireless Sensor Networks: Mobile wireless sensor networks are the wireless sensor networks with movable nodes. More illustratively, contrary to traditional WSN, which is generally static in nature, nodes are able move from one place to another. This movement may further reduce their longevity, but it facilitates a lot in the applications viz. wildlife monitoring etc. Also, the nodes' movement results in a frequently changing network topology, restricting the use of existing solutions developed for a traditional WSNs in MWSNs. Such node mobility not only causes the dynamically changing network topology but also brings the intermittent connectivity among the nodes before the researchers' community. Moreover, like other variants of OppNets, MWSNs allow two nodes to exchange their packets only when they come in close contact with each other.

Despite adding many design complexities, MWSNs are contributing a lot to humanity via a number of applications such as wildlife tracking etc.

Applications and Projects Based on Opportunistic Computing

Opportunistic networks, popularly known as OppNets, have evolved from traditional MANETs and have been popular enough in the researcher community and also in industry due to enormous possibilities they represent. As explained above, in opportunistic networks, nodes communicate only when they come into contact with one another, similar to the way humans communicate with one another; hence, they are referred to as human-centric. Since their inception, OppNets have been used to provide connectivity in emergency scenarios where the fixed network infrastructure would have been destroyed due to disaster. They are used to provide network services in rural and suburban areas where limited infrastructure exists due to low economic investments. Many battlefield applications have been developed using OppNets. Its human-centric nature has also enabled a number of social applications such as mobile social networking, social discovery, content sharing, and interaction supports for users with common interests or social relationships. By collecting information from the neighbor devices held by other nearby users, and through the use of various sensors (cameras, GPS transceivers, accelerometer etc.) mounted on the smartphones being carried by users, OppNets can be deployed for pervasive and urban sensing as well. Nowadays, OppNets are also being used for offloading cellular networks. Many such applications are regularly being deployed using OppNets which are directly or indirectly benefiting human society; some such applications are discussed in more detail as follows.

Disaster Management

In emergency situations where the existing network infrastructure may get destroyed as a result of disaster, traditional networks fail to provide services for the first responders. However, in such unplanned and unexpected scenarios, a quick response may save many lives. In such emergency scenarios, opportunistic networks play a very significant role as they require no infrastructure; nodes in the network support the store-carry-forward strategy as explained above; and, most importantly routes are created dynamically between the source and destination, suiting the needs of emergencies. Opportunistic networks aim to provide communication services to struggling segments of the devastated telecom infrastructure to achieve the aforementioned noble objective. They not only connect the surviving parts of the networks, but also provide a considerably solid platform to deploy other essential services/applications. Most of the time in emergency situations, the rescue personnel are equipped with mobile devices such as smart phones, PDAs, etc. to be used for different purposes like triage and tracking of victims. Some solution/ projects which have been successfully deployed in such disastrous situations are listed below.

Haggle Electronic Triage Tag: Haggle Electronic Triage Tag, popularly known as Haggle-ETT, is a project developed and deployed for emergency situations; it utilizes opportunistic communication (Martin-Campillo et al., 2010). It uses Haggle, which is an open-source architecture for opportunistic and data-oriented networking, along with mobile devices to generate the desirable Electronic Triage Tags (ETTs).

Tactical Medical Coordination System: This is a project popularly known as TacMedCS which has been deployed as a military system to evaluate and show the real-time data of casualties occurred in the battlefield (Williams, 2007). The project involves handheld devices which capture the casualty data and the respective GPS information; the captured information is then transported for the further treatment via satellite communication. Moreover, to achieve a collaborative computation, different handheld devices are made connected through IEEE 802.11 mesh communication.

The Mobile Agent Electronic Triage Tag System: This project was developed to be deployed in emergency situations (Marti et al., 2009). The system comprises mobile agents which store and carry the triage information about the disaster victim. Afterwards, the mobile agents are provisioned to move through mobile ad hoc networks to forward the stored information for appropriate decision-making process. Moreover, the system utilizes the concept of time-to-return (TTR) for the mobile agent migration decision.

Pervasive Healthcare Services

Opportunistic computing when combined with existing network technologies may result in a more promising and pervasive network of intelligent devices like sensors and actuators which are installed over a patient's body to generate various measurements at different levels.

Intelligent Transportation System

With the evolution of opportunistic computing, vehicular ad hoc networks have been seen as the most vital example of OppNets; they demonstrate not only vehicle-to-vehicle communication but also the communication between vehicles and fixed RSUs in order to improve the traffic efficiency and commuters' safety. Other than the main objective of providing road safety, several other applications, like broadcasting weather conditions, parking assistance, obstacle alerts, vision improvement, route finding, location finding, entertainment for the commuters, etc., have also been found as the outcomes of VANETs. Some of the exemplary applications of the vehicular ad hoc networks are as follows:

SmartPark: This is a very popular application of VANETs which was proposed in 2005 (SmartPark, 2005). With the help of sensors mounted on the vehicles and preinstalled RSUs, the application collects data for parking space availability and guides the vehicle's driver to a free parking spot. The application involves assigning each parking spot a wireless mote to verify its occupancy; the signal sent from these wireless motes are received by the sensor mounted on the vehicles to the confirm the free parking space.

CarTorrent: Similar to the BitTorrent, a peer-to-peer file sharing system with the name CarTorrent (Nandan et al., 2005; Lee et al., 2007) has been developed for vehicular networks and features a parallel download from multiple hosts simultaneously. The peer-to-peer content sharing has been successfully portrayed even in the presence of several design constraints that come with vehicular ad hoc networks.

Remote patient monitoring: An application has been developed for remote medical monitoring to collect the physiological data of patients who might be out of the coverage area of the hospital premises (Noshadi et al., 2008). The application utilizes VANETs for patients' data collection via sensors mounted on the vehicles visiting the patients' places of residence. This approach has proved its significance where traditional communication networks have failed. Not only the patients' data but also environmental measurements like noise level, pollution level, etc. are collected to help make medical decisions for the improvement of patients via this approach.

Cellular Traffic Offloading

With the evolution of 4G technology, the number of cellular users has increased worldwide. With this increased cellular population, demands for the real-time services like audio/video streaming etc. have also increased drastically. Especially during popular events like cricket matches or music concerts, simultaneous demands for Internet data may degrade the overall performance of local antennas being shared in providing the services to temporal crowd. In an attempt to satisfy such temporary needs and to prevent the serving antenna from being overloaded, some local caching points may be installed to provide the content of common interest to the concerned users. The same concept is proposed in an application namely Mobile Advanced Delivery Server (MADServer) (Petz et al., 2012). In his proposal, Petz et al. suggested a middleware be installed in between a web server and a web application. MADServer proposes the time-critical data be sent using a traditional cellular approach while the time-insensitive data is sent opportunistically; however, content categorization in the context of user-interest is still an open issue to be resolved.

Applications for Challenged Environments

As stated earlier, traditional network technologies have failed in situations where there is neither any wired and wireless network infrastructure nor any dependable and reliable energy supplies. To meet the need of transporting data from such environments, some additional concepts are added, like data mules and gateways. A data mule is a moving entity, such as buses, boats, trackers, etc., and the gateway is a network entity which is connected to the Internet or the data center with strong and constant connectivity. The data mules act as an intermediary. They are used to carry the requests to the gateway from the challenged environments; they also carry the responses from the gateway to the challenged environments. Several projects have already been proposed in this regard, with a few being mentioned below.

Bytewala Project: The project was developed and deployed to provide Internet services to remote places with poor or no networking infrastructure in Africa (Bytewala, 2013). In this project, travelers act like data mules in such a way that when they travel to some place with better network infrastructure, they carry requests and while returning back to their originating places, they carry the responses. The entire approach has been possible using DTN2 reference implementation.

BeachNet: BeachNet is a project for content sharing in a network scenario with variable numbers of static nodes (Ott et al., 2011). Through the use of opportunistic networks, data pertaining to the beach area or picnic places are disseminated via the store-carry-forward strategy.

DakNet: The DakNet project has been widely deployed in remote places and rural areas in India and Cambodia to provide a platform for e-governance and e-marketing. The idea in this project is to enable poor and distant communities to avail of government services and to use the information regarding goods' market. Here, the vehicles—buses, bikes, carts, etc.—are treated as data mules.

Forwarding in Opportunistic Networking

As discussed in detail above, the primary requirement for message routing in the traditional network is the establishment of an end-to-end path between the two communicating hosts. Once the routes are confirmed, the message is transported for its time-bound delivery to the designated recipient. Any occurrence of delay or disruption is treated as a routing failure. However, this may not always be the case, especially in applications like deep space communication, underwater communication, emergency scenarios, etc. As discussed earlier, in such applications, data transmission takes place whenever a suitable contact is found between the hosts. Since the occurrence of contacts are unpredictable and nodes are completely unaware of any such happening, routes between the hosts are developed dynamically and any node can act as a next hop in the communication, provided the node is more likely to bring the message nearer to the destination. Moreover, as the routes are not identified by the transmitting node (to the destination) before sending the message as in the traditional networks, the term "forwarding" is found more suitable in the place of "routing" in opportunistic computing.

Forwarding of data in opportunistic networks is more challenging as compared to its traditional counterparts. Moreover, the more fundamental question in message forwarding in opportunistic network is to figure out the node to which message can be transported, with an objective of maximizing message delivery and minimizing the delay. A number of forwarding solutions have already been defined by the researcher community, each differing in the degree of knowledge or contextual information. Here, the context of the user is defined as current system state surrounding the user. More illustratively, context of the user can be defined in terms of location, temperature, etc. (referred to as physical context); in terms of the available resources and services (i.e., system attributes); and in terms of preferences, habits, etc. of the user (i.e., user context) (Dey, 2001). The available set of opportunistic forwarding protocols is classified into three major categories on the basis of amount of contextual information being utilized in the solution. They are *no context, partial context,* and *full context* forwarding protocols.

No context forwarding protocols transport the message in the network without any prior knowledge; *partial context* forwarding protocols utilize the network information, such as number and status of the neighborhood nodes and previous contact history etc., in the forwarding of the message; and, lastly, *full context* forwarding protocols decide to whom to forward the message—not only the basis of network attributes but also the status of the surrounding environment plays a very critical role in deciding the next node in communication. Thus, the full context forwarding protocols have the maximum computational complexity. In the later subsections, some examples of each category will be discussed briefly.

No Context Forwarding Protocols

Protocols in this category use neither any network information like neighborhood status and network topology nor any surrounding environmental details in deciding the message forwarding process, i.e., forwarding decisions are taken only on the basis of the nodes' own status. Some of the instances of this category are listed as follows.

Epidemic

As implied by the name, the epidemic forwarding protocol believes in disseminating the message like spreading a disease (Vahdat and Becker, 2000). More illustratively, the message to be transported is forwarded to each of its neighborhood nodes by the sender and, in turn, the receiving nodes repeat the same process. Thus, multiple copies of a single message are made to float in the network with an objective of maximizing the delivery chances while minimizing the delay. The epidemic approach has been found efficient in message delivery and in delay, but the excessive message floating increasing in proportion to the number of nodes in network becomes a network overhead.

Single-Copy Algorithms

Contrary to the epidemic forwarding protocol, in which multiple copies of the same message are flooded in the network, as implied by the name itself, only a single copy of the message is forwarded in the network under the single-copy algorithms (Spyropoulous et al., 2004). A set of protocols has been proposed under the umbrella of single-copy algorithms, suiting different application requirements, as follows:

Direct transmission: In direct transmission, a message is forwarded only once when the contact is made to the destination. Though the approach may result in a huge delay, only a single copy of the message is transmitted in the network.

Randomized routing algorithm: This approach forwards the message with a probability *p* towards the destination.

Utility-based routing: In this approach, a utility function is defined to compute the significance of the candidate nodes on the basis of number of meeting among a pair of nodes and time elapsed since the last encounter of the nodes.

Seek and focus routing protocol: The seek and focus routing protocol first follows the randomized routing protocol and then switches to utility-based routing protocol once a highly significant node is confirmed through the utility function.

Spray and Wait Forwarding

In the spray and wait forwarding protocol, *n* copies of the message (to be forwarded to the destinations) are created by the sender which are handed over to *n* distinct neighborhood nodes of sender. These *n* neighbors then carry the message to the destination using the direct transmission strategy of the aforementioned single-copy algorithms.

Partial Context Forwarding Protocols

Partial context forwarding protocols require the network information to decide to whom to forward the message. Network information pertaining to this approach may be neighborhood status, contact history, network topology, node mobility, etc. All those protocols which require any of the aforementioned information can be considered as belonging to partial context forwarding protocols. Some of the most popular instances of this category are listed below with brief explanations.

PRoPHET

In an attempt to minimize the overhead of epidemic forwarding protocol, Probabilistic Routing Protocol using History of Encounters and Transitivity, popularly known as PRoPHET, is proposed (Lindgren et al., 2003). PRoPHET utilizes the contact history and mobility pattern of the nodes in deciding the nodes' candidature in the message forwarding process. It defines and uses a probabilistic metric called the *delivery probability*, with the values ranging between *0* and *1* which simply determines the likelihood of message delivery to the intended destination. In the employed probability model, mobility patterns of the nodes are identified, with an assumption that the node mobility is not random, and if the nodes have frequently visited some particular field portions, then there is a high probability of visiting the same portion in the near future.

Context-Aware Routing (CAR)

Like PRoPHET, context-aware routing (CAR) requires some network information a priori for its functioning. The required information are the neighborhood degree and nodes' current energy status. CAR tries to figure out if the destination belongs to the same connected part of the sender and, if so, it forwards the message using a more popular MANET routing protocol: Destination sequenced distance vector (DSDV) routing protocol. Otherwise, a utility function is called based upon the change rate of connectivity and the probability of the destination being in the same cluster of the relay node.

HYMAD

HYMAD is a hybrid of the MANET and DTN routing protocols (Whitbeck and Conan, 2010). HYMAD stimulates the nodes to identify the groups of connected neighbors and forward the messages using MANET routing protocols (use of the MANET routing protocol within the group is due to the fact that they perform well with a low overhead among the nodes of close vicinity) within the groups; however, the communication among the different groups is facilitated using DTN references.

Full Context Forwarding Protocols

Protocols in this category utilize the surrounding information along with that of network. They may also require the information like users' habits and their preferences. Some of the more popular forwarding protocols under this categorization are listed below.

SimBet

This protocol is based on the social interactions, utilizing two different metrics—centrality and social similarity—to determine the probability of the node with which it is expected to contact the intended destination (Daly and Haahr, 2007)

The centrality metric is used derive the message exchange among the communities, as the nodes with high centrality values will act like bridges between different communities in the network. On the other hand, the other parameter—the similarity metric—is used to deliver the message to the intended destination within the community, as the nodes with higher values of similarity measures imply higher probabilities of finding a common neighbor to the destination.

HIBOp

HIBOp is a history-based forwarding protocol for opportunistic networks which employs the knowledge of network topology, contact history of the nodes, and user context to enable nodes to learn their similarity with the others. HIBOp provisions a message to be spread among similar users with a belief that the node with similarity information to that of the destination has a higher probability of meeting the destination. For its functioning, HIBOp asks the users to share their information like their home and work addresses, hobbies, preferences, etc.

In addition to the aforementioned forwarding protocols, several other protocols have been proposed by researchers addressing various specific applications in opportunistic networks, viz. BubbleRap (Hui et al., 2011) SSAR (Li et al., 2010), PeopleRank (Mtibaa et al., 2010), SREP (Xie et al., 2011) and 3R (Vu et al., 2011). Although all such protocols, along with those in previously mentioned two categories—No Context and Partial Context—have addressed the problem of message forwarding in different aspects, still there are some open issues, like energy consumption (especially in PSNs), deployment of a large-scale OppNets, social behaviour of the users, etc.

References

I. F. Akyildiz, D. Pompili and T. Melodia, Underwater acoustic sensor networks: Research challenges. *Ad Hoc Networks*, vol. 3, pp. 257–279, 2005.

S. Burleigh, A. Hooke, L. Torgerson, K. Fall, V. Cerf, B. Durst, K. Scott and H. Weiss, Delay-tolerant networking: An approach to interplanetary internet. *IEEE Communications Magazine*, vol. 41, no. 6, pp. 128–136, 2003.

Bytewala, Bytewala project, 2013. Available at: http://www.tslab.ssvl.kth.se/csd/project s/092106/.

E. Daly and M. Haahr, Social network analysis for routing in disconnected delay tolerant MANETs. In: *Proceedings of the Eighth ACM International Symposium on Mobile Ad Hoc Networking and Computing*, ACM, 2007, pp. 32–40.

A. Dey, Understanding and using context. *Personal and Ubiquitous Computing*, vol. 5, no. 1, pp. 4–7, 2001.

K. Fall, A delay-tolerant network architecture for challenged internets. In: *Proceedings of the 2003 Conference on Applications, Technologies, Architectures, and Protocols for Computer Communications*, 2003, pp. 27–34.

P. Hui, J. Crowcroft and E. Yoneki, Bubble rap: Social-based forwarding in delay tolerant networks. *IEEE Transactions on Mobile Computing*, vol. 10, no. 11, pp. 1576–1589, 2011.

K. Lee, S.-H. Lee, R. Cheung, U. Lee and M. Gerla, First experience with cartorrent in a real vehicular ad hoc network testbed. In: *Mobile Networking for Vehicular Environments*, 2007, pp. 109–114.

Q. Li, S. Zhu and G. Cao, Routing in socially selfish delay tolerant networks. In: *Proceedings IEEE INFOCOM*, 2010, pp. 1–9.

A. Lindgren, A. Doria, and O. Schelen. Probabilistic routing in intermittently connected networks. *ACM SIGMOBILE Mobile Computing and Communications Review*, vol. 7, no. 3, pp. 19–20, 2003.

R. Martí, S. Robles, A. Martín-Campillo and J. Cucurull, Providing early resource allocation during emergencies: The mobile triage tag. *Journal of Network and Computer Applications*, vol. 32, no. 6, pp. 1167–1182, 2009.

A. Martín-Campillo, J. Crow Croft, E. Yoneki, R. Martí and C. Martínez-García, Using Haggle to create an electronic triage tag. In: *The Second International Workshop on Mobile Opportunistic Networking—ACM/SIGMOBILE MobiOpp*, 2010, pp. 167–170.

V. F. S. Mota, F. D. Cunha, D. F. Macedo, J. M. S. Nogueira and A. A. F. Loureiro, Protocols, mobility models and tools in opportunistic networks: A survey. *Computer Communications*, vol. 48, pp. 5–19, 2014.

A. Mtibaa, M. May, C. Diot and M. Ammar, Peoplerank: Social opportunistic forwarding. In: *Proceedings IEEE INFOCOM*, 2010, pp. 1–5.

A. Nandan, S. Das, G. Pau, M. Gerla and M. Sanadidi, Co-operative downloading in vehicular ad-hoc wireless networks. In: *2nd Annual Conference on Wireless On-demand Network Systems and Services, WONS*, 2005, pp. 32–41.

H. Noshadi, E. Giordano, H. Hagopian, G. Pau, M. Gerla and M. Sarrafzadeh, Remotemedical monitoring through vehicular ad hoc network. In: *IEEE 68th Vehicular Technology Conference, VTC*, 2008, pp. 1–5.

J. Ott, A. Keränen and E. Hyytiä, Beachnet: Propagation-based information sharing in mostly static networks. In: *Proceedings of the Third Extreme Conference on Communication. The Amazon Expedition, ExtremeCom '11*, 2011, pp. 15:1–15:6.

A. Petz, A. Lindgren, P. Hui and C. Julien, Madserver: A server architecture for mobile advanced delivery. In: *Proceedings of the Seventh ACM International Workshop on Challenged Networks*, 2012, pp. 17–22.

Project SmartPark, Parking made easy, 2005. Available at: http://smartpark.epfl.ch.

T. Spyropoulos, K. Psounis and C. Raghavendra, Single-copy routing in intermittently connected mobile networks. In: *First Annual IEEE Communications Society Conference on Sensor and Ad Hoc Communications and Networks*, 2004, pp. 235–244.

A. Vahdat and D. Becker, Epidemic routing for partially connected ad hoc networks. Tech. Report, Duke University, 2000.

L. Vu, Q. Do and K. Nahrstedt, 3R: Fine-grained encounter-based routing in delay tolerant networks. In: *IEEE International Symposium on a World of Wireless, Mobile and Multimedia Networks (WoWMoM)*, 2011, pp. 1–6.

J. Whitbeck and V. Conan, Hymad: Hybrid DTN-MANET routing for dense and highly dynamic wireless networks. *Computer Communications*, vol. 33, no. 13, pp. 1483–1492, 2010.

D. Williams, Tactical Medical Coordination System (TacMedCS), Naval Health Research Center. San Diego, California. Technical Report, February 2004–June 2007; November 2007.

X. Xie, Y. Zhang, C. Dai and M. Song, Social relationship enhanced predicable routing in opportunistic network. In: *Seventh International Conference on Mobile Ad-hoc and Sensor Networks (MSN)*, 2011, pp. 268–275.

Index

Printed in the United States
by Baker & Taylor Publisher Services